International Conference on Sustainable Cities (ICSC 2016)

E3S Web of Conferences Volume 6 (2016)

Ekaterinburg, Russia
19 May 2016

Editors:

S. Kortov
I. Balk
N. Terlyga

ISBN: 978-1-5108-2606-9

Printed from e-media with permission by:

Curran Associates, Inc.
57 Morehouse Lane
Red Hook, NY 12571

Some format issues inherent in the e-media version may also appear in this print version.

Printed by Curran Associates, Inc. (2016)

For additional information, please contact EDP Sciences – Web of Conferences
at the address below.

EDP Sciences – Web of Conferences
17, Avenue du Hoggar
Parc d'Activité de Courtabœuf
BP 112
F-91944 Les Ulis Cedex A
France

Phone: +33 (0) 1 69 18 75 75
Fax: +33 (0) 1 69 28 84 91

contact-edps@webofconferences.org

Additional copies of this publication are available from:

Curran Associates, Inc.
57 Morehouse Lane
Red Hook, NY 12571 USA
Phone: 845-758-0400
Fax: 845-758-2633
Email: curran@proceedings.com
Web: www.proceedings.com

TABLE OF CONTENTS

E3S Web of Conferences
Open access conference proceedings
in Environment, Energy & Earth Sciences
A flexible, fast and efficient
publication service,
for an open access recording
of your event
and a wide dissemination of
your proceedings
www.e3s-conferences.org
edp open Open Access Publishing
from EDP Sciences

Aims & scope

E3S Web of Conferences is an open-access journal devoted to the publishing of conference proceedings dealing with Environment, Energy and Earth Sciences.

Coverage includes technological and scientific aspects as well as social and economic matters. Major disciplines include: soil science, hydrology, oceanography, climatology, geology, geography, energy engineering (production, distribution and storage), renewable energy, sustainable development, natural resources management, etc.

The journal aims to make conference proceedings visible, citable and widely usable by the scientific community. It offers outstanding services for the publication and dissemination of contents in order to ensure long-term archiving and maximum exposure of scientific events.

Why publish with E3S Web of Conferences?

- **Open access:** conference papers available worldwide.
- **Fast publication:** less than 3 months.
- **Total flexibility:** no page restrictions, free unlimited use of colour in the online version, any kind of scientific document welcomed (traditional articles, posters, abstracts and slideshows).
- **High visibility:** all documents indexed in main databases and easily found online.
- **Fair author rights:** CC-BY Creative Commons license.
- **Range of additional publishing options:** printed proceedings, CD-ROMs, copy-editing, etc.

Contact: Isaline Augusto
e3s@webofconferences.org

ISSN (electronic edition): 2267-1242

www.e3s-conferences.org

edp open Open Access Publishing from EDP Sciences

Statement of Peer review

In submitting conference proceedings to E3S Web of Conferences, I certify to the Publisher that I adhere to the **Policy on Publishing Integrity** of the journal in order to safeguard good scientific practice in publishing.

1. All articles have been subjected to peer review administered by the proceedings editors.
2. Reviews have been conducted by expert referees, who have been requested to provide unbiased and constructive comments aimed, whenever possible, at improving the work.
3. Proceedings editors have taken all reasonable steps to ensure the quality of the materials they publish and their decision to accept or reject a paper for publication has been based only on the merits of the work and the relevance to the journal.

Title, date and place of the conference

International conference on sustainable cities
May 19, 2016
In Yekaterinburg, in Yeltsin Centre

Proceedings editor(s):

Sergei Kortov – First Vice-Rector, Ural Federal University
Nadezhda Tarlyga – Deputy First Vice-Rector, Ural Federal University
Igor Balk – Managing Member, Global Innovation Labs (USA)

Date and editor's signature

1.07.2016.

17, avenue du Hoggar - PA de Courtabœuf – BP 112 - 91944 Les Ulis Cedex A (France)
Tél. : 33 (0)1 69 18 75 75 - Fax : 33(0)1 69 07 45 17 – www.edpsciences.org

E3S Web of Conferences 6, 01001 (2016) DOI: 10.1051/e3sconf/20160601001
ICSC 2016

The model of localized business community economic development under limited financial resources: computer model and experiment

Dmitry Berg[1a], *Rustam* Davletbaev[1], *Olga* Zvereva[1], and *Dmitry* Nodjenko[2]

[1]Ural Federal University, 620002, Yekaterinburg, Russia
[2]Institute of Economics, Ural Branch of RAS, 620014, Yekaterinburg, Russia

Abstract. Globalization processes now affect and are affected by most of organizations, different type resources, and the natural environment. One of the main restrictions initiated by these processes is the financial one: money turnover in global markets leads to its concentration in the certain financial centers, and local business communities suffer from the money lack. This work discusses the advantages of complementary currency introduction into a local economics. By the computer simulation with the engineered program model and the real economic experiment it was proved that the complementary currency does not compete with the traditional currency, furthermore, it acts in compliance with it, providing conditions for the sustainable business community development.

1 Introduction

Globalization processes, which concepts were established only in the 70s of the last century, now affect and are affected by business and work organizations, economics, socio-cultural resources, and the natural environment. Large-scale globalization began in the nineteenth century [1] and was accompanied both by positive and negative phenomena in trade and transactions, capital and investment movements, migration and movement of people [2]. Real threats of globalization stimulated the anti-globalist movement and development of the glocalization vision (glocalization is a portmanteau of globalization and localization). The last one provides city-to-city cooperation; encouraging peace building and international development in the non-governmental sector. It emphasizes the central role of cities in international relations and the role of local authorities in the world governance system [3, 4].

One of the main restrictions, the local business community is faced by, is the financial one: money turnover in global markets leads to its concentration in the certain financial centers, which are far away from local communities. Local economic activity is inhibited because of the money lack: economic communications between agents (their goods and products exchange) are hampered and becomes more difficult. This problem arises all over the world and local communities try to overcome it using such instruments as bills

[a] Corresponding author: Bergd@mail.ru

of exchange, barter systems, local exchange trading systems (LETS), different types of complementary currencies, and etc. [5].

The main goal of the present study is to demonstrate the positive effect of complementary currency usage in the local economic turnover which is provided for sustainable development of the local business community. The effect proposed is shown in computer simulation using the local community communications agent-based program model and is also confirmed by the real economic experiment.

2 Local community communications model

To prove the concepts discussed above communications program model was engineered. In the modeled system several manufacturing entities have to produce some goods and in order to succeed in this activity try to supply themselves with resources in the specified volumes.

To design this model agent-based technology was used [6]. This technology is geared towards social sciences and is the most suitable for this problem solving [7]. It supports bottom-up approach: one should model agents, their interactions with each other and environment and observe the whole system parameters (macroparameters) as a result.

Agent's attributes were calculated on the base of Leonief's static equilibrium (Input-Output model) [8]. The Model was designed in Netlogo [9], free software framework, which supports the agent-based technology. Model window is shown in Figure 1.

In the Model there are N agents, every agent produces a unique product. The *i-th* agent product volume is denoted as x_i. The ordered sequence of all agent product volumes (x_i, $i = \overline{1, N}$) forms ($\vec{X} = (x_i)_{i=1}^{N}$), which is the production vector of the economic system under study.

The *i-th* agent consumes the other agent's products for manufacturing of his own product of value x_i. These consumption volumes are denoted as w_{ki}, where w_{ki} indicates the *i-th* agent consumption of the *k-th* agent's product. Then $\vec{W_i}$ is the *i-th* agent consumption vector:

$$\vec{W_i} = (w_{ik})_{k=1}^{N}, \tag{1}$$

this vector lists the agent product volumes to be consumed by the *i-th* agent in order to manufacture his own product of value x_i.

Every consumption volume (w_{ki}) is in direct proportion to x_i and is determined by the *i-th* agent productive technology. Technical (input) coefficients estimated as $a_{ki} = \frac{w_{ki}}{x_i}$ are dimensionless. These coefficients a_{ki} ($k=\overline{1, N}$, $i=\overline{1, N}$) comprise structural matrix, or table (A_{NxN}). Matrix element a_{ki} determines the *k-th* agent's product volume which is consumed by the *i-th* agent for its single product unit manufacturing.

Then the *i-th* agent consumption vector can be expressed as:

$$\vec{W_i} = [a_{1i} \cdot x_i, a_{2i} \cdot x_i, \dots, a_{ni} \cdot x_i], \tag{2}$$

after product exchanges every agent has a remaining volume of his own product. Partly it can be used in his manufacturing process (this is reflected in $\vec{W_i}$ and the structural matrix) and the last part denoted as y_i can be spent to meet the internal non-productive needs. Vector $\vec{Y}$ is known as a vector of final demands:

$$\vec{Y} = (y_i)_{i=1}^{N}, \tag{3}$$

for the *i-th* agent it is true that:

$$x_i = a_{1i} \cdot x_1 + a_{2i} \cdot x_2 + \ldots + a_{Ni} \cdot x_N + y_i. \quad (4)$$

Static form of Leontief's Equilibrium (SLE) is settled by the following vector equation:

$$\vec{X} - A\vec{X} = \vec{Y}, \quad (5)$$

Leontief's static model has several limitations, one of them – this model describes a closed system. It means that all products are manufactured and consumed inside the system. This could be considered to be unrealistic. Modern economics is a global economics: a local economic system can't supply itself with all the necessary goods and install relations in the import-export form with other local communities.

Thus, the special agent was introduced into the model. This agent represents the environment in the system, per se it accumulates all the requirements in external resources, and all the shipments for external consumers. This agent, becoming the part of the system, corresponds to the requirements of SLE.

To support exchange operations in the model money was introduced. There are two types of currency in the system: external (real) currency and internal (complementary) currency. The external currency is money in its common sense. It can be used in all exchange operations simulated in the model and complementary currency is used only for internal exchanges, between all but the environmental agent.

Every agent has these currencies in its account, their volumes are in the direct proportion to its program and are calculated with the supplement coefficients *k_mon* (for real currency) and *k_vmon* (for complementary currency). These coefficients can be set by special controls in the model window.

Simulation period is divided into cycles of exchange. In every cycle of exchange every agent receives an opportunity to buy one kind of resource according to its consumption vector. In the model five different exchange strategies are realized. In details they are discussed in [10, 11]. In this experimental series the first "list" strategy was under research. In accordance to this strategy an agent is looking through its consumption list from the start to the end and tries to find and meet the first non-zero requirement in resources in the maximal possible value. After all the agents make their attempts, model time increases and new exchange cycle is initiated. Simulation lasts until any exchange is possible, i.e. there is a *j-th* agent (P_j), who has non-zero *k-th* value in the current consumption list ($w_{kj} \neq 0$) and non-zero sum of money, and, at the same time, there is a *k-th* agent has some product for sale (current value of $x_k \neq 0$).

$$[\exists P_j, \exists P_k: (w_{kj} \neq 0) \wedge (m_j \neq 0) \wedge (x_k \neq 0)], \quad (6)$$

The rule of money spending in exchange operation depends on the agent type and was already discussed.

Experiments were held using different data sets. These data sets are represented in Table 1.

Table 1. Data sets used in experiments

Data Set	Structural Matrix (A)
1	$\forall i = 1..20, \forall j = 1..20, \quad a_{ij} = 0.0385$
2	$\forall i = 1..20, \forall k = 1..20, \sum_{i=1}^{20} a_{ki} = \sum_{i=1}^{20} a_{ik} = 0.77$
3	$\begin{cases} \forall k = 1..20: \sum_{i=1}^{20} a_{ki} = \sum_{i=1}^{20} a_{ik} = 0.77 \\ \forall m = 1..20, \forall n = 1..20, a_{mn} = a_{nm} \end{cases}$

In the first dataset all agents have equal consumption demands for products of the other agents, every structural matrix *(A)* element is equal to 0,385. The second dataset is made for the agents that have equal total demands for the other agents' products. And, in the third dataset, besides equal total agents' demands the condition of mutual demand equality is true (i.e. every *i-th* agent tries to consume the *k-th* agent product in the same volume as the *k-th* agent tries to consume the *i-th* agent product).

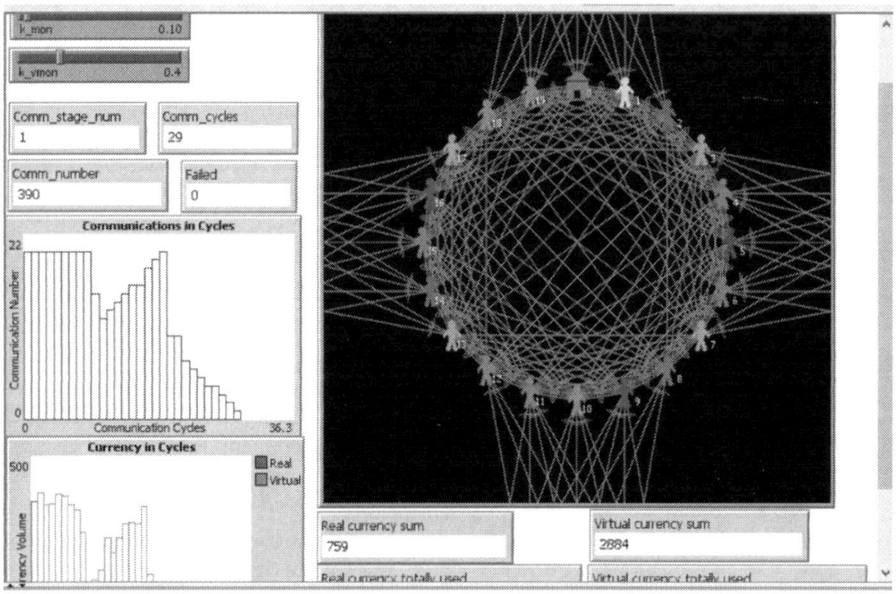

Fig. 1. Model window.

The experiments have proved:

- that if the model is based on the static Leontief's Equilibrium all exchanges are fulfilled successfully in all money conditions (if at least k_mon >0);
- complementary currency existence in the system greatly decreases the time necessary for exchange completion, the complementary money "works" as the real money (it is apparent from Figure 2b);
- complementary currency weakens, and when it is enough, suppress the crisis (Figure 2a) which arises if money volume is insufficient in the system;
- the less real currency volume is in the system, the greater positive effect is made by the complementary currency introducing into the system (Figure 3).

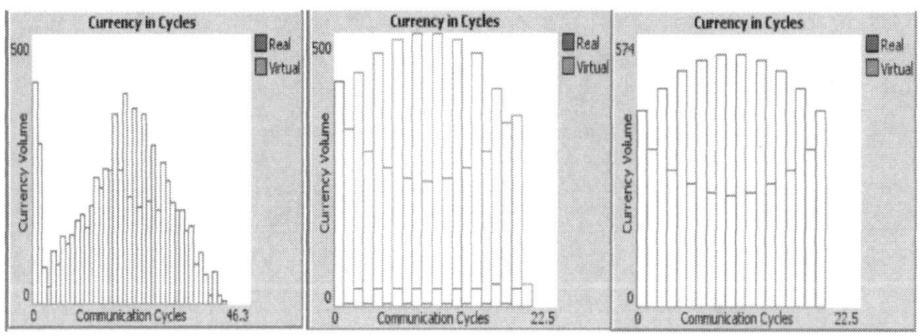

a) $k_{mon}=0.1;\ k_{vmon}=0$ *b)* $k_{mon}=0.1;\ k_{vmon}=0.4$ *c)* $k_{mon}=1.0;\ k_{vmon}=0$

Fig. 2. Money turnover in the system for the second data set: external (real) currency (red), internal or virtual (complementary) currency (green).

3 Economic experiments in the local business community

To carry out the economic experiment a local business community from the agricultural region of Bashkortostan republic was chosen. This region is in 40 km far from Ufa, there are several small manufacturing firms and service sector enterprises in the five closely located villages. Agricultural output and products manufactured from it comprise the economic basis of the community. The aggregated indicators of the community are as follows:

- the number of employees is under 200 persons;
- the business activity types are as follows: crop farming, cattle breeding, trading, agricultural machinery services;
- the area of cultivated land is 8000 hectares, the area of arable land is 6000 hectares;
- the cattle number is 1200;
- the level of capitalization is 330 million rubles.

At the beginning of the experiment, the community was in a depressive state, as bank loan debts were growing, there were almost no cash for salary payment and the salary debt was greater than for three months, thus, all of this resulted in the social tension increase.

In the period of 2010–2013 three economic experiments on commodity exchange were carried out in the territory of the community on base of the complementary currency usage. The complementary currency was fully supported by the locally manufactured product and trade residues in local shops.

The largest local enterprise of the community, "Shaymuratovo Ltd.", has become an emission center of complementary currency. It issued the local currency (product checks) in the cash form, this currency was named "shaimuratics". Every year product checks of the previous issue were withdrawn from the turnover and replaced by the new ones. The annual issue of the complementary currency was in the ranks of 100 thousand to 400 thousand rubles.

There were no restrictions on the complementary currency usage: it was used for salary payments as well as for mutual payments between entrepreneurs. The complementary currency usage was a freewill case: every community member in any day had a possibility to receive the necessary currency volume (no more than his/her salary amount). That is why "shaimuratics" were in demand in the case when there was a lack of ruble cash for supporting the turnover. Ruble and complementary currency turnover control was executed when they were paid (in the enterprises cash-desks) and when they were received as payments for products and services (in the shops cash-desks, and etc.). All data about the complementary currency and rubles turnover was aggregated for a week. It gave a possibility to estimate how many rubles and "shaimuratics" were involved in the turnover for a week. The turnover diagram is presented in Figure 3. Turnover scales are made comparable (look at the left and right range) for better visual illustrating of revealed synergetic effect which arose due to the currencies joint use.

It is evident from Figure 3 that it took at about 5 weeks "to launch" the turnover, during this period the community members were studying to use the new type currency. Later on, the turnover demonstrated the tendency towards the growth, it was true for rubles and also for the local currency. At the same time, in the neighbor communities which used only rubles the turnover was continuously decreasing. The week level of turnover has the strongly marked season characteristic: the 29th – 32nd week peak became the result of the New Year eve, then one can see the drop (the 33rd week) because of activity decrease while the winter vacations.

From the diagram one can comprehend that in the periods when there was the lack of rubles in the community, it was compensated by the complementary currency, and vice versa, when the rubles volume was large enough, the necessity in the complementary currency became less. Thus, complementary currency in a local business community acts as an addition and in compliance with the national (real) currency.

While the experiment, the complementary currency usage provided sustainable development of the local business community according to the following parameters:

- commodity turnover has been increased twelve times;
- productivity has been increased by 20 percent;
- all salary debts were repaid;
- average salary level has been increased by 26 percents.

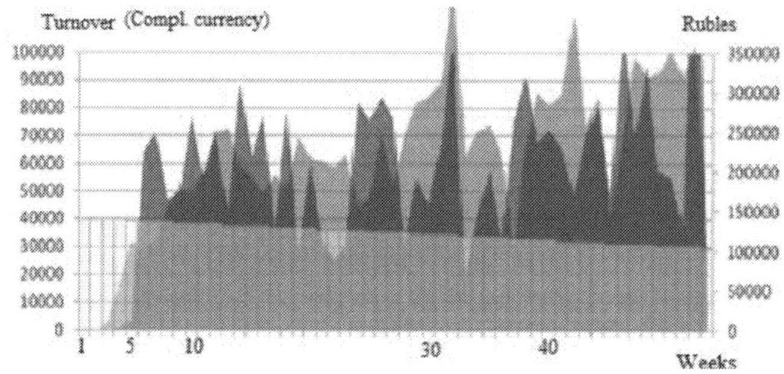

Fig. 3. Comparable dynamics of the local community commodity turnover of the complementary currency (is shown in blue color, left range) and Russian rubles (is shown in red color, right range) in terms of week specification (Data was collected in 2010–2011).

Conclusion

The two types experimental results (in the computer simulation and in the real economic experiment) show that introduction of complementary currency removes the restrictions arisen in the local business communities because of the cash outflows to financial centers of globalization and, thus, results in creation of the conditions for sustainable development of regions.

The mechanisms of the complementary currency, which positively influence the community economics, are as follows. Although there are money outflows to financial centers of globalization, there are capital goods, labor force, local resources and consumer demands in the local territory. But the consumer demand occurs to be a potential one, because there are no instruments of payment (cash). Complementary currency helps to solve this problem, it enables local products purchasing, local enterprises get the possibility to increase production, can exchange with local providers, and also begin to pay salaries in complementary currency. Later on, this reproduction cycle is repeated. A lot of enterprises due to product sale for complementary currency can operate above the break-even point, which is impossible in the conditions of payments in the national cash.

The synergetic effect of simultaneous currencies usage was revealed: complementary currency does not compete with the national currency, it acts in compliance with it providing the conditions for sustainable development of a community as a whole.

Acknowledgements

Present study was carried out under financial support of the Russian Fund of Fundamental Research grant № 15-06-04863 "Mathematical models of local payment system lifecycles".

References

1. K.H. O'Rourke, J.G. Williamson, NBER Working Paper Series, **7632** (April 2000). http://www.nber.org/papers/w7632.pdf
2. International Monetary Fund. IMF Publications (April 2000). http://www.imf.org/external/np/exr/ib/2000/041200to.htm
3. P. Guinness, *Geography for the IB Diploma Global Interactions* (Cambridge, 2011)
4. I.G. Nordin, C. Zamorano Llena (eds), *Redefinitions of Irish Identity: A Postnationalist Approach* (Oxford, 2010)
5. B.A. Lietar, *The Future of Money. Creating New Wealth, Work and Wiser World* (Moscow, 2007)
6. C.M. Macal, M.J. North, Journal of Simulation, **4**, 151–162 (2010)
7. V.V. Leontief, *Essays in economics. Theories, theorizing, facts and policies* (Moscow, 1990)
8. P.L. Borrill, L. Tesfatsion, *Agent-Based Modeling: The Right Mathematics for the Social Sciences. Working Paper № 10023* (July, 2010)
9. NetLogo. http://ccl.northwestern.edu/netlogo
10. O.M. Zvereva, D.B. Berg, *OPT-i 2014 - 1st International Conference on Engineering and Applied Sciences Optimization, Proceedings*, 33–43 (2014)
11. D.B. Berg, O.M. Zvereva, A.G. Shelomentsev, A. Taubayev, *Proceedings 15th International Multidisciplinary Scientific Geoconference SGEM 2015. Ecology, Economics, Education and Legislation*, Vol. 3, 109–118 (2015)

Research of ecology-friendly technology for elimination of ammonium from waste water through the use of floating bed

Nikolay Makisha[1a]

[1]Moscow State University of Civil Engineering, 129337, Moscow, Russia

Abstract. The article touches upon integrated treatment of sewage. It reveals description of the research with use of floating feed. Experiment data were the basis to prove effectiveness of floating feed as a way to intensify treatment processes in aeration tanks. The article also includes math treatment of the results to correspond them with laws of enzymatic kinetics.

Introduction

Nitrification/denitrification researches have a long history and strong background in viewpoint of process dependencies and influence of factors of the environment [1]. There are also a large number of technological solutions and constructions towards removal of ammonium compounds out of wastewater by means of aeration tanks as well as biofilters. However there is still a wide scope of work to be done [2].

Among these options treatment quality improvement by means of immobilized biomass seems to be one of the most prospective. Researchers worldwide have already invented various types of materials as well as the technologies to implement them. Polymer materials have certain advantages to be applied for these needs [3, 4].

Researchers of Moscow state university of civil engineering also have a solid base of knowledge in the field of waste water treatment in general and ways how to force it in particular. One of the recent researches was a complicated study aimed at deep ammonium removal by means of floating bed. There are several aims to be pursued: to estimate how a floating bed influences the treatment process, amount of floating bed to be involved, what the treatment duration is etc. [5–8].

Research description

The research was divided into several stages to set and solve abovementioned tasks consequently [9]. Stage 1 had two subtasks. The first one was to define if a use of floating bed gives any advantage in viewpoint of deep biological treatment. The second subtask was to estimate the concentration of floating bed in the tank that allows to obtain the most efficient results. There were three samples of polymer bed taken for the research. Each of

[a] Corresponding author: nmakisha@gmail.com

them had a different shape: parallelepiped – type 1 (Figure 1), cube – type 2 (Figure 2) and sphere type 3 (Figure 3); different properties such as volume weight and specific surface (see table 1); and different material of production: polystyrene and polyethylene [10–12].

Table 1. Properties of researched bed

Type of bed	Volume weight, kg/m^3	Specific surface, m^2/m^3
Type 1	22	420
Type 2	25	600
Type 3	50	620

Samples of floating bed were investigated in the laboratory model in which a traditional technological scheme of biological treatment by means of aeration tank was implemented (Figure 4). Each type of bed was investigated with 10%, 20% and 30% ratio of bed volume to tank volume. Each type of bed has been simultaneously investigated for each three ratios above. A control tank (without bed) has also been installed to verify the results obtained. Treatment duration was 8 hours.

Fig. 1. Floating bed, type 1.

Fig. 2. Floating bed, type 2.

Fig. 3. Floating bed, type 3.

Research at each stage has been carried out by means of artificial waste water that was by made on peptone basis. Concentration of pollutants for such waste water fully corresponded to common domestic waste water. Sanitary analyses were performed according to acknowledged methods [13–15].

Efficiency of ammonium removal in the tanks with floating bed was constantly higher than in control tank through the whole stage 1. That fact speaks in favour of positive influence of floating bed on processes of biological treatment of waste water. The optimum ratio between bed and tank volumes was determined on the basis of laboratory experiments for each type of bed and further mathematical processing of practical data. Load on sludge

 DOI: 10.1051/e3sconf/20160601002

and velocity of oxidation for both BOD and NH4 were calculated. The math description of the dependence between activity of enzyme reaction and substrate concentration was given according to the Michaelis – Menten equation and graphical representation of Lineweaver – Burk [9, 11].

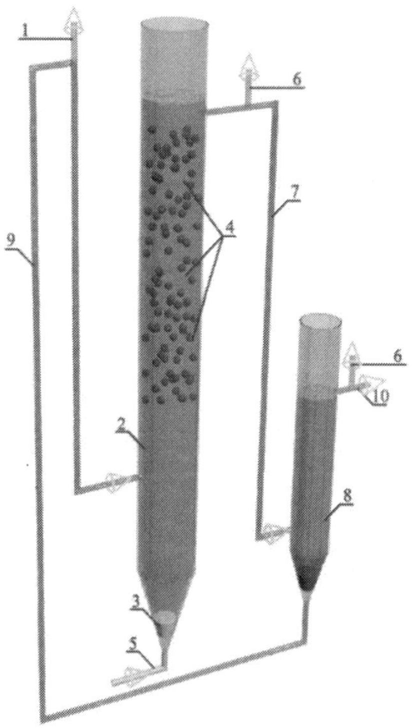

Fig. 4. Laboratory model, stage 1: 1 – inlet; 2 – aeration tank; 3 – aerator; 4 – floating bed; 5 – air inlet; 6 – compressor; 7 – inlet of sludge mixture to secondary clarifier; 8 – secondary clarifier; 9 – recirculation of active sludge; 10 – discharge of treated water.

Floating bed of type 1 showed its best results for research aims with a ratio of 10%, meanwhile for the floating beds of type 2 and 3 the optimum was 20% of tank volume. Table 2 shows average values of inlet and outlet pollutants' concentrations (BOD and NH_4) for models with optimum values of ratio above.

Table 2. Average results for stage 1

Type of bed	Average values, mg/l					
	BOD_5			NH_4		
	In	Out	E,%	In	Out	E,%
Type 1	155	8	95	22,1	3,1	86
Type 2	161	6,7	96	26,6	4,9	82
Type 3	144	6,4	96	22,7	4,3	81
Control	150	12,6	92	23,0	10,3	55

Thus, experiments helped to achieve a proof that treatment (organic and NH_4 pollution) processes have been intensified through the use of floating beds of various types. Despite of the more or less positive results, discharge concentrations especially for NH_4 were still far from regulation requirements, so further investigation had to be carried out. The first alteration happened was a substitution technological scheme. It was enhanced by the introduction of biofiltration reactor on the second step of treatment while the same aeration reactor was used at the first step. There were three laboratory models under simultaneous investigation – each with its' own type of floating bed installed. Biofiltration tanks has the same bed for each models – cylindrical polymer bed with fibers of polyethylene on its' surface [12].

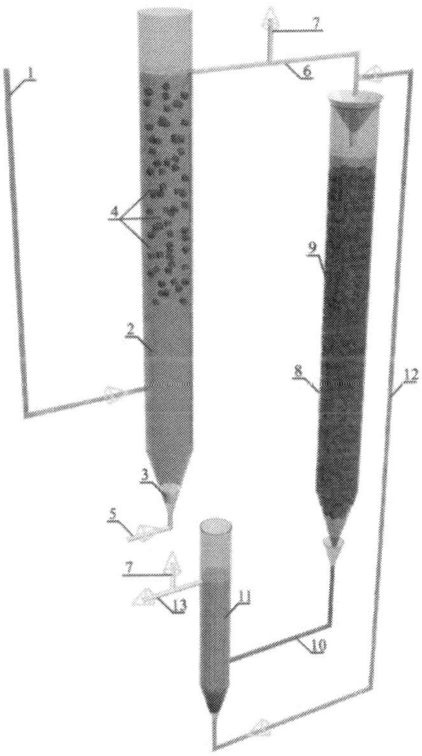

Fig. 5. Laboratory model, stage 2: 1 – inlet; 2 – aeration reactor; 3 – aerator; 4 – floating bed; 5 – air inlet; 6 – compressor; 7 – outlet of treated water to biofiltration reactor; 8 – biofiltration reactor; 9 – fixed bed of biofiltration reactor; 10 – discharge of water to secondary clarifier; 11 – secondary clarifier; 12 – recirculation of active sludge; 13 – discharge of treated water.

Once the optimum ratios had been determined on the stage 1 all further researches were made on this basis. Through the whole stage all three types of feed were working simultaneously. Oxygen concentration was at a level of 7 mg/l during the whole stage 2. It was marked that aeration reactor works in a more stable way if no recirculation flow I provided. Concentrations of free and immobilized sludge were near 0.2 mg/l and 0.9 mg/l correspondingly. After the overview of the analyses results it was defined, that type 2 showed more efficient treatment (see table 3). The entire efficiency of system grew; however, the regulation requirements were not still met for nitrates and nitrites.

Table 3. Average results for stage 2

Type of bed	Average values, mg/l					
	BOD_5			NH_4		
	In	Out	E,%	In	Out	E,%
Type 1	162	31	81	23,5	6,7	74
Type 2	164	14,6	91	21,3	2,7	89
Type 3	171	21,1	88	25	14	40

Math processing of the results allowed obtaining the values of velocity of oxidation and denitrification. The Michaelis – Menten equation can be written as follows:

$$V = \frac{V_{max} \cdot S}{k_m + S} = \frac{47.6 \cdot L_{ex}}{3.3 + L_{ex}} \quad (1)$$

$$V = \frac{V_{max} \cdot S}{k_m + S} = \frac{14.3 \cdot NH_{4ex}}{4 + NH_{4ex}} \quad (2)$$

The investigation towards enhacement of removal of organic pollution and nitrogen compounds were continued at the stage 3. The two-step scheme of treatment was still in use, however two steps changed their places – biofiltration reactor was on the first step, aeration reactor – on the second (Figure 6). There was also intermediate clarifier installed after the biofiltration reactor to hold the excessive biofilm inside. That measure helped to keep relatively low sludge concentration inside aeration reactor that was quite efficient on the previous stage.

Table 4. Average results for stage 3

Type of bed	Average values, mg/l					
	BOD_5			NH_4		
	In	Out	E,%	In	Out	E,%
Type 1	49,3	15,7	68	14,1	2,3	83
Type 2	39,4	9,5	76	27	17	37
Type 3	49,6	3.0	92	11.3	0,4	97

The highest treatment efficiency about 99% on the stage 3 was obtained by the floating bed type 3 that made concentrations of BOD, NH_4, NO_2 and NO_3 meeting the regulation requirements (see table 4). Oxygen concentrations remained at the same level – 6–7 mg/l, sludge concentrations for free floating and fix phase were 0.8 mg/l and 0.3 mg/l correspondingly.

Math processing of the results allowed obtaining the values of velocity of oxidation and denitrification. The Michaelis – Menten equation can be written as follows:

$$V = \frac{V_{max} \cdot S}{k_m + S} = \frac{18.9 \cdot L_{ex}}{4.17 + L_{ex}} \quad (3)$$

$$V = \frac{V_{max} \cdot S}{k_m + S} = \frac{5.56 \cdot NH_{4ex}}{0.67 + NH_{4ex}} \quad (4)$$

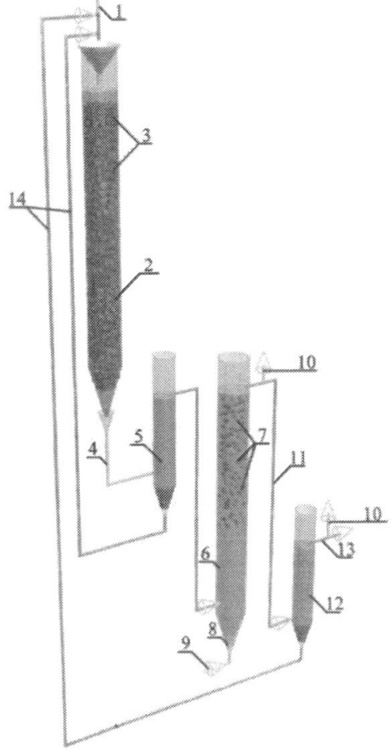

Fig. 6. Laboratory model, stage 3: 1 – inlet; 2 – biofiltration reactor; 3 – fixed bed of biofiltration reactor; 4 – discharge of water to intermediate clarifier; 5 – intermediate clarifier; 6 – compressor; 7 – floating bed; 8 – aerator; 9 – air inlet; 10 – air discharge: 11 – discharge of water to secondary clarifier; 12 – secondary clarifier; 13 – discharge of treated water; 14 – recirculation of active sludge.

Conclusion

Experiments proved confirmed that use of floating bed with ratio of 20% of bed volume to aeration tank volume enhances the efficiency of biological treatment of waste water and removal of organic pollution and nitrogen compounds

The research validated the stability of biological system with the use of immobilized sludge and under the concentration of free floatinf sludge below 0.8 mg/l.

Low sludge concentration in the aeration tank results in low excessive sludge production that reduces the cost of sludge treatment to make the whole treatment process more energy efficient.

References

1. V.A. Vavilin, *Duration of biomass exchange and destruction of organic compound in systems of biological treatment* (1986)
2. N.A. Zaletova, *Urban wastewater treatment from biogenic substances: nitrogen and phosphorus*: the dissertation … of doctor of technical sciences (1999)
3. N. Makisha, V. Scherbakov, A. Smirnov, E. Scherbina, IJAER, **10**, 24, 44347–44349 (2015)

4. V. Scherbakov, E. Gogina, T. Schukina, N. Kuznetsova, N. Makisha, E. Poupyrev, IJAER. **10**, 21, 44353–44356 (2015)
5. E. Gogina, I. Gulshin, Pr. Eng., **117**, 107–113 (2015)
6. I. Gulshin, A. Kuzina, IJAER, **10**, 21, 42618–42623 (2015)
7. E.S. Gogina, O.V. Yantsen, O.A. Ruzhitskaya, AMM,**580–583**, 2354–2357 (2014)
8. O.A. Ruzhitskaya, E.S. Gogina, AMR, **919–921**, 2141–2144 (2014)
9. N. Makisha, E. Gogina, AMM, **361–363**, 632–635 (2013)
10. O.A. Ruzhitskaya, E.S. Gogina, AMR, **919–921**, 2153–2156 (2014)
11. Elena Gogina, Nikolay Makisha, AMM, **587–589**, 636–639 (2014)
12. N. Makisha, O. Yantsen, AMM, **587–589**, 640–643 (2014)
13. N. Makisha, E. Gogina, AMM, **587-589**, 644–647 (2014)
14. N. Makisha, Y. Voronov, E. Poupyrev, V. Volshanik, IJAER, **10**, 21, 41919–41922 (2015)
15. N. Zaletova, Y. Voronov, N. Makisha, IJAER, **10**, 21, 42454–42455 (2015)

Humanization of the Russian housing model in the postmodern space

Leyla Leonova[1a], *Elena* Alpatova[1], *Stefan* Kowalczyk[2], and *Tatyana* Rutkauskas[1]

[1]Ural State University, 620002, Yekaterinburg, Russia
[2] President Livia Projects International Inc., Calgary, Alberta, Canada

Abstract. The article analyzes the existing approaches of management in the sphere of housing and communal service in Russia (HCS). They have not improved the quality of life and have not ensured sustainable reproduction of the human capital. According to the authors, the implementation in practice of Anglo-Saxon governance model in Russia has not proved the historical context of the country's development. A comparative analysis was conducted of housing sphere of Russia and the European Union. It is recommended to use the Scandinavian management model as a more acceptable for the Russian conditions.

Signs of global post-industrial era have undergone changes: digital technologies and social experiments have become symbols of modernity. Technological advances are taking place in the world. However, a significant part of the population is not provided by basic housing infrastructure. They live in bad conditions. They exposed to detrimental environmental factors. In the past two decades the social inequality has deepened, the "unauthorized" settlement of immigrants in large cities has appeared. EU countries also have fully felt the "tsunami" of refugees and migrants in 2014–2015.

The growth of destructive socio-political trends devalues people's desire for a "better" life.

Economic, political and environmental turbulence in system leads to the realization of "fragility" and doom the future, deprive each person of faith in a positive future world order. Prigogine believed that «the world evolving from order to chaos» [1]. Chaos is not the final state of the universe. The new structures are born from it. However, the transition states themselves cause the people to feel fear and rejection.

In the context of the cultural space of postmodernity the apprehension of frightening multivariance progress has come. It brings not only technological advances, but also wars, destruction and alienation. It is difficult to see a reference point of civilization to the man of post-industrial society.

On the other hand, postmodernity has brought the idea of a permanent social change to the world, change incentives for the formation of the inner freedom of the individual. Postmodern, released from Europe, covered all socio-cultural and economic models of the reflecting world. The dominant postmodern trend has reduced the dependence of the individual from the institutions of politics and economics.

[a] Corresponding author: lel.leo@mail.ru

The life of modern cities is a typical example of postmodern culture with economic imbalances and social segregation: "Intelligent" high-rise buildings of infill development literally "are pushed" on land plots between dilapidated buildings today, digital and physical world grow into each other, national, demographic composition of people has uncontrolled changes. There are the questions arise about the "protection" of life: settlements are controlled by electronic systems, but they also reveal their vulnerability [2].

Civilization general models of urban management develop In the context of postmodern. We have acquired new technologies, methods and mechanisms of interaction, for a whole century, however we have lost the value targets.

An excursion into the history of the issue allows us to analyze the changes of management in housing and communal services (HCS) since the beginning of the XX century to the present and pay attention to aspects of modern management which require immediate changes in management approaches.

Basic management methodology in housing and communal sphere in Russia had been included in the publication of L.A. Velikhov, "Fundamentals of Municipal Economy" in 1928 [3].

Of course, a person is not regarded in the system as the central link, but he is a means, an "element" of the socialist city. A secondary role was assigned for comfortable housing, infrastructure and services sector.

However, the ideology of the urban economy had a social entity. This ideology has been adopted as the basis in the Soviet period of management. Housing and communal sphere has been recognized as acute social, necessary for the reproduction of the population. The start of mass housing construction in the USSR began in 1954. For fifteen years (to 1969), 1.469 billion square meters of housing was built [4]. Due to the famous "Khrushchev" people moved from dugouts, communal houses to separate apartments. At the same time, all the engineering support of infrastructure systems in the houses has been established. This infrastructure has remained viable until the '90s.

In the so-called "of reform" period (since 1991), HCS was based on the life-support facilities of the Soviet period. They quickly grow old. It required large financial costs of subsidizing sphere (major repairs of residential buildings, urban relaying of utilities, construction of pumping stations, sewage treatment plants, waste recycling). However, the country's attention was focused on the interesting political events of "perestroika".

Since the adoption of the law "On privatization of housing stock in the RSFSR" in 1991 [5] and transition of HCS to a break-even operating mode marked departure from the elements of the social nature of the system, despite the fact that the poor families needed the targeted social protection. Often, because of bureaucratic obstacles the assistance was a great problem. The system of subsidies for socially disadvantaged groups of population exists since today.

Russian HCS is a quality of life indicator which shows the critical depreciation of fixed assets, in 2014 it was about 43–52%. For example, the wear of water supply and sanitation systems in some Russian cities reaches 80–90% [6]. According to experts, the wear of equipment more than 50% exceeds the threshold economic security indicators in the country in 2 times. Table 1 shows the wear of the basic structure of HCS funds [6].

Table 1. Depreciation of fixed assets in the construction, production and distribution of water, gas and electricity, %

Type of economic activity	2008	2009	2010	2011	2012	2013	2014
All fixed assets	43,5	45,2	45,3	45,3	47,1	47,9	48,6
Manufacture and distribution of electricity, gas and water	55,6	52,2	51,2	50,7	51,1	51,2	51,2
Building	42,3	44,6	45,5	46,9	48,3	49,4	47,3

Other community and social services	42,9	39,8	40,7	42,5	44,5	45,1	43,4

Poor infrastructure, high level of corruption and growing social inequality among the population are the results of the conservative modernization during the Soviet and post-Soviet period. More than 4.5 million Russian citizens live in dilapidated and emergency housing. 40% of buildings (homes) need major repairing [7].

More than ten years ago monography "Why Russia is not America" (A. Parshev) [8] and a series of articles S. Kara-Murza "Problems in the Russian house" [9] were published. Later the "King-cold, or why Russia freezes" (S. Kara-Murza) was published too. [10]. The authors of the articles carry out the idea of the distinctive features of the housing and communal sector in Russia.

Only poor population of this country is not able to pay for all housing difficulties. This point of view is also relevant today.

We believe that in the second decade of the twenty-first century a deep change in management conception of housing and communal services in Russia is necessary. The traditional features of the national economy also must be taken into account. What is the base of this position? Firstly, housing and communal sphere in Russia is not only the production of market services. This area is strongly connected with the political system of the state and is always played as a "trump card" during the every election political campaign.

Secondly, there are the whole monopoly in production and delivery of life-support resources and the absence of a healthy competition in HCS in Russia. In the sphere, where monopoly firms dictate the rules, there is not any basis for competition and the formation of the healthy market relations. And this is, in turn, leads to a constant increase in tariffs.

Thirdly, for the formation of the healthy market relations in the housing and communal sector a system of cooperation between all market counterparties is necessary. This interaction based on eleven process organization principles. They are presented in Table 2.

Table 2. Basic principles of market relations in Russia's sphere of housing and communal services

The principle of a market economy	Summary principle	Features housing and communal sector
Freedom of choice	The right market subject to purchase and use their own resources	Companies – suppliers of resources – monopolists. The customer has no right to select either a resource or a supplier
Competition	Availability filled independently functioning market sellers and buyers of products and services	Sellers of resources are mainly monopolists. Consumers do not have the right to choose a supplier The centralization of systems engineering house life support
Market segmentation	Market Share in the consumer groups according to income Identifying market segments that are "served" by competitors	There is no segmentation in apartment buildings Consumers with high incomes purchase apartments of larger area
Targeting	The poor families need the targeted social protection.	Rigid frame subsidies (Payment of housing services should be more than 22% of total family income). Bureaucracy of the subsidy process
Paid	The obligation of timely and fully payment for the supplied resources and rendered	A large number of non-payers because of low incomes and inflated prices

	services	
Free market pricing	Prices are formed on the basis of market conditions	Regulation of prices and tariffs by regional energy commissions Government regulation of limited indices of prices and tariffs increase. Management companies have the right to adjust the prices and tariffs for the provision of certain types of works and services
Contractual relationship	Market entities enter into contracts for the provision of utility services There is a consumer protection	The owners of apartment buildings have neither the right nor the ability to enter into separate contracts for the supply of resources. Consumers are not protected from the low quality services
Government regulation	Enforcement of existing legislation by market participants	Inadequate of housing law and legislation in the housing and communal sphere
Formation of different standards of housing preferences	Consumer preferences opportunities for choice of place of residence	New and modern apartments are about 25% of the housing. Low income of 70% of the population does not allow them to buy it
Consumer priority	The responsibility of manufacturers and retailers to consumers	The consumer have the risk of receiving low quality services
Interchangeability of works (services)	The consumer has the right to replace inappropriate services	This principle is impossible because of centralizing infrastructure of life support systems

In this article we compared the main market interaction principles and features of housing and communal sector and came to the conclusion that all of them do not promote effective governance in market relations. And it means that we could not blindly adopt market mechanisms and principals in the field of non-market relations. Besides, we can see a long influence of persistent negative factors to Russian economy.

Could the urban metabolism system keep the maintain stability under the negative influence factors? What management model will be suitable in the structure of non-stationary economic heritage processes? To answer these questions, the authors appealed to the international experience of housing and communal management. EU development strategy up to 2020: the reduction of the number of people at risk of poverty or social exclusion is the exhibitor of humanitarian progress in Europe. The poorest segment of population was identified by comparing the range of incomes and living conditions of the population [11]. This index is composed from the analysis of three indicators: the degree of poverty and deprivation and the number of people who have problems with job.

To sum up, the housing and communal sphere is socially oriented in all European countries. State authorities and governments kept under control HCS in all EU countries. Municipalities regulate the work of private enterprises in HCS and defend the rights of citizens. Municipalities guarantee a high quality of services. Sweden, Finland, Germany, Poland, England: the municipal management is a business in the form of management companies. Along with them the owners of apartments control housing maintenance. They come together into the non-profit consumer organizations.

Associations of tenants and homeowners provide the working of the apartment complex. They dispose of financial resources and buy necessary services. The process of their activity is similar to the work of functioning of HOA (homeowners) in Russia.

On one hand, we can see the tight control from the public and the state organizations. On the other hand – there is the formation of a society clear position in the communal conflicts relations. This protects the residents from fraud and corruption.

The studying of positive foreign experience in apartment management has a considerable interest for Russians, because about 85% of them are living in apartment buildings. Table 3 shows features of apartment management models in EU (compiled by the authors).

Table 3. Features of apartment (housing) management models in EU countries (compiled by the authors)

Country	Features of management model in apartment houses	Result
Sweden	An extensive network of housing unions and associations promotes the rights of tenants and owners; have the political authority The work of the housing court Differ payment service models: - "discount model" – the selection and payment of actual services. - "Depreciation model" – reduction of fees in case of correct treatment of the subject property	Ensuring clarity in the activities of control structures and rates, detection and suppression of corruption
Finland	Form of housing management – Management companies 70% are controlled by Management companies, 30% – Council House Management companies make operational control in accordance with the contract The contract is a list of services, prices and conditions Management company selects the operating organizations to provide communal services Operating organizations choose a method of providing services: 1) independently 2) enter into contracts with specialized companies (repair and construction, landscaping, protection, waste recycling). The operator make annual report on costs and incomes. In the big cities of engineering support systems are in municipal ownership. Heating and water supplying systems are centralized. Power plants may be in privately owner	Minimization of costs and the application of advanced technologies (For example, the presence of oil reserve stations during low temperature (-30 °C), using of waste heat homes)
Germany	Apartment house is managed independently The amount of payment is determined by the general meeting. Housemasters make cleaning. Each owner of the apartment enters into a contract with their own resources, supplier – water, electricity, gas. Individual heating, due to boiler	Full independence of the residents, cost minimization
England	The existence of private companies – suppliers of resources. Each resident chooses resource provider, if it is necessary, can change them, but not more than once a week	Independent search vendor minimizes resource rent
Poland	Joint-stock company with 100%-ing state (municipal) capital. All kinds of public services combined into a single state-owned company. She manages the finances, production resources and monitors activities. Resource Companies may have a different form of property	Providing quality services. Ensuring the reliability of housing and communal services

The apartment buildings management forms abroad are very different: from the conclusion of individual or collective contracts with inhabitants of an apartment house with different resource providers as in England and Germany to the total municipal capital management in Poland.

But in spite of the management forms the main feature is to minimize the cost of consumers. And as a result we can see the rent reduction, clarity and transparency of costs and funding in HCS. This makes virtually impossible the existence of corruption.

HCS is very important in Scandinavia countries. Tenants' Association has a significant influence on public policy. In Finland, for example, as well as in Russia, there is a central heating and water supplying systems.

We have analyzed the work of the urban and municipal services of the above mentioned countries and selected the basic management models:

- "Anglo-Saxon" – provides total autonomy management to every inhabitant. – "Scandinavian" – complete control of governments. Vital public services and resources such as heating and water supplying are centrally delivered. There is the "double" law protection of consumers. Residents are protected by the court and the government.
- "German" is individual management of each apartment house. Municipals control the autonomous engineering supplying systems.

In the Russian HCS sequentially implanted Anglo-Saxon governance model. However, the market management model does not show its effectiveness in the modern Russian conditions. The real market in Russia is impossible because of its historical heritage. The fact is a natural monopoly of the main suppliers of vital resources, such as gas, electricity, heat.

Russian HCS is not only an economic problem. This is an important political aspect of Russia's life, and is an extremely important social factor not only because of consumer behavior, but also the factor of surviving and reproduction of the human capital in harsh Russian cold climate.

For example, 85% of the population has a monthly income per person about one-two living wages In Russia. According to the Federal Law "On Living Wage" [12] the average per capita income does not exceed 8500 rubles. The monthly fee apartment for a family of three people is about 7000 rubles.

The government should regulate prices because prices are rising by 8% a year. But it must also to create an effective legal and legislative basis in the housing and communal sphere.

The rate of growth of tariffs and costs are quite different from those in the EU. For example, from 2001 to 2009, the prices increased by 6 times in Russia. Meanwhile the inflation rate in Russia was lower and prices for HCS in the European countries increased no more than 50% [13]. Table 4 shows the calculated data of significant increase trend of tariffs for HCS from 2001 to 2010 in terms of average oil price of $ 54.83 per barrel as the main source of livelihood in Russia [14]

Table 4. Increase of tariffs for HCS in 2009 in real terms, % from the 2001 [15]

Services	Russia	EU	Germany	Poland	Finland	Great Britain
HCS and fuel	144,5	12,0	5,5	19,2	15,7	25,2
Rent	–	-0,7	-4,4	14,7	10,6.	3,1
The content of the housing	150,4		6,3	5,2	0,8	8,6
Water supplying	199,2		-1,2	39,3	14,9	31,8
Electricity	62,7		32,0	23,1	33,2	54,6
Gas	98,6		16,6	45,8	–	108,4

Heat supplying	172,5		22,9	9,3	45,3	–

In the 90 years of the twentieth century, Russia was in a situation of duality contrasting models: socialist and market. According to the authors, unconditional transition to market concepts without taking into account national features and specificities was originally risky for the main areas of the economy, including HCS sector.

Postmodern as a system of flexible features, offers us "intermediate" model for countries with ethnic, cultural and historical characteristics. And these models are successfully operating. Models balance between the economic freedom principle and social regulation.

Having the positive examples of the European experience, Russian housing reformers installation to the "Anglo-Saxon" management model could not be considered as the only appropriate in Russia. It is based on the total privatization of vital important infrastructure and the total market economy transition.

Because of its cold climate and mentality of the population, as well as limited resources and the level of living of its poor population, the most acceptable for Russia could be the "Scandinavian model" as a model of "intermediate" type.

Taking into account the historical monopolism and centralization in HCS, as well as the current state of the economy it is necessary to use systematic public and government monitoring of the HCS management in Russia. It is necessary to take into account the historically established monopolism, centralization of essential housing and communal services, and the current state of the economy in Russia. Public authorities, government and society should control the management of housing and communal services.

References

1. I. Prigogine, I. Stengers, *Order out of chaos. Man's new dialogue with nature* (London 1984)
2. G. O`Tuathail, The Postmodern Geopolitical Condition: States, Statecraft, and Security at the Millennium Annals of the Association of American Geographers, **90-1** (2000)
3. L.A. Velikhov, *Fundamentals of urban management* (1996)
4. History of housing and communal services in Russia, http://www.muromraion.ru/node/3652
5. Federal Law of 04.07.1991 No. 1541-1 (ed. of 16.10.2012) "On privatization of housing stock in the Russian Federation", http://www.consultant.ru
6. B. Livshits, *System analysis of market reform unsteady economy of Russia: 1992–2013* (2013)
7. International experience in communal sphere, http://dgkh.mos.ru/international-experience-of-housing
8. A. Parshev, *Why Russia is not America?* (2006)
9. S. Kara-Murza, S. Telegin, *Problems in the Russian house* (2004)
10. S. Kara-Murza, S. Telegin, *King-cold freezes or why Russia* (2006)
11. Household composition statistics, http://ec.europa.eu/eurostat/statistics-explained/index.php/Household_composition_statistics
12. Federal Law of 24.10.1997 No. 134 "On Living Wage", http://www.garant.ru
13. Europe in figures – Eurostat yearbook 2012: Living conditions and social protection Eurostat (2013)
14. The dynamics of oil prices since 1990, http:/tass.ru/economika/1572991
15. O.N. Kraeva, Scientific electronic journal, http://uecs.ru/uecs49-492013/item/1941-2013-01-21-11-01-42

Urban management: Addressing the housing and utility challenge

Oksana Nikitina[1a], *Elena* Alpatova[1], *Elena* Shershneva[1], and *Konstantin* Rutkauskas[1]

[1]Ural State University, 620002, Yekaterinburg, Russia

Abstract. The authors have studied the three aspects of urban management: apartment buildings, major repair management, requirements for facility management companies to be licensed and waste management. The paper presents all the mentioned above aspects and possible negative circumstances and proposes solutions to the problems.

Urban development is due to the potential of accumulated professional experience, resources, communication. When resources are depleted, the experts begin to think about changing strategic priorities. Large cities have a strong attraction to the life currents of civilization. The rapid development of urban space means the formation of a special "spirit" of the agglomeration, filled with contra

The driver of urban growth was the "boom" of housing in metropolitan areas. For example, more than 1 million square meters were erected of new housing each year in the Yekaterinburg. However, geopolitical events have led today to the classical "crisis of overproduction" and the rise of the imbalance "demand – offer" [1]. Falling demand market can not completely "absorb" large-scale urban development projects.

In this context, it is time to shift the focus from programs "increase in square meters" on the quality of urban management procedures that would ensure the survival of the urban economy in a period of economic fluctuations.

During the 2014–2015 three approaches to the transformation of urban economy management have been developed. The first innovation is associated with the financial support of the "heavy" direction of Housing and Communal Services (HCS) – overhaul of buildings [2]. The second innovation concerns the scope of control and supervision over the management control of companies (CC) [3]. The third innovation is the new concept in the cleaning of household waste [4].

Innovation 1. Introduction fee overhaul. Based on the analysis of the features of municipal services sphere management authors have identified trends in the development of management innovations and proposed measures for their improvement.

At the beginning of 2016, there were 15 cities with "millionaire" in Russia [5]. This is metropolis with its social, cultural, ecological features. Along with intelligent architectural forms, creative design projects landscapes of modern cities today clouded view of houses is not a presentable appearance. The average specific gravity of the problem in the general fund housing volume is shown in Table 1. As a part of the problem fund – dilapidated and emergency housing.

[a] Corresponding author: elen-kon@yandex.ru

Dilapidated housing is not only disfigures architectural image of the city, but also requires the cost of maintaining accident-free "status." The experts found the consequences more dramatic order. Miserable habitat destruction is the health, social relations and personality. This is a very dangerous type of destruction: it quickly becomes irreversible. Bio-psycho-social disorders appear in the structure of personality.

The man, who sinks to the bottom of life, almost can not restore the previous level of welfare.

Table 1. The share of problematic housing in the total housing stock of the Russian Federation, as a percentage [5]

Regions	2010	2012	2013	2014
Russian Federation	3,08	3	2,8	2,69
Central Federal District	1,92	1,8	1,73	1,66
Moscow	0,33	0,4	0,38	0,39
Northwestern Federal District	3,28	3,2	2,6	2,36
Saint Petersburg	0,65	0,4	0,39	0,02
Southern Federal District (from 2010)	2	1,8	1,74	1,59
North Caucasian Federal District	6,75	6,5	6,59	6,54
Volga Federal District	2,48	2,5	2,43	2,41
Ural Federal District	3,09	3,3	3	3,06
Kurgan region	5,73	5,7	5,47	5,5
Sverdlovsk region	1,86	1,9	1,97	2
Tyumen region	5,75	6,4	5,32	5,57
Khanty-Mansi Autonomous District	6,96	7,3	6,88	7,42
Yamalo-Nenets Autonomous District	8,6	11,7	10,19	10,41
Chelyabinsk region	1,62	1,6	1,56	1,47
Siberian Federal District	4,74	4,4	3,95	3,7
Far Eastern Federal District	6,12	5,9	5,67	5,86

There comes a habit of bad forms of existence. Man falls into the cycle of "dependence" of poor living circumstances of which can not get out on their own.

Since 2014 apartment buildings expected to be repaired by the fund overhaul (FO) or accounts opened with homeowners' associations (HOA). Regional operators are authorized to dispose of these accounts [6].

FO due to the "pocket" of taxpayers and for purpose of expenses can become a full participant in the financial market by analogy with the pension, health, social security funds. Finance management system overhaul consists of two parts: income generation and distribution costs.

Income generation system has three "stumbling block" in the way of practical implementation of the reforms conceived.

1. The process of accumulation of funds for repairs. The principle of "common pot" is contrary to the Civil Code and the Constitution of Russia. Those who have a home in good condition contain their property and other people's property: the funds collected from tenants of a house can go for another overhaul of the strategic plan of repairs. Setting priorities are not standardized for the regions and is "random" nature.

2. Weak control over the formation of the of payment size for overhaul. Governments set limits on the amount of monthly payment to the Major Repair Fund.

Table 2. The cost of major repairs to 1 sq. meter of housing area in rubles [7]

Regions	2014	2015
Moscow	7,30	15,00
Saint Petersburg	2,00	2,00–3,00
Ural Federal District		
Kurgan region	6,97	6,97
Sverdlovsk region	6,10	8,20
Tyumen region	5.60–20,11	7,50
Khanty-Mansi Autonomous District	8,55–13,85	9,10–14,7
Yamalo-Nenets Autonomous District	10,50	10,50

In urban regions fee overhaul varies greatly. For example, St. Petersburg has a minimal fee. The city is known monuments they repaired at the expense of the federal budget. The minimum rate set in the Sverdlovsk region for 2015–2016 amounts to 8 rubles per sq. m (Yekaterinburg is 8.52 rubles). This is 23.25% more than the level of the tariff for 2014 [8, 9].

History of the problem shows that the owners always pay for the overhaul. Only positions in settlement documents from time to time changed the name. For example, in Yekaterinburg in 2003 the position of "Overhaul" was included in "Payment for the maintenance of housing" (1.65 rubles per sq. m). Since March 2004 position "Payment for the maintenance of housing" became known as the "Maintenance" (The rate increased 1.46 times compared to 2003). In early 2008, the position "Maintenance" has teamed up with the position of "property of Contents" and has increased 1.94 times. Merge and rename articles of utility bills has continued since 2004. Thus, the "latent" requisitions for repair turned into millions of rubles, and the public is not exposed the cost of supervision, timing and quality of repairs made. The current situation shows the next fiscal "difficulties" of the program – regional operator of Sverdlovsk region managed to complete major repairs only 30% of the planned number of houses [10].

In some European countries the fee overhaul was introduced. Various tariff plans have been used. Thus, the inhabitants of one city, living in different houses, pay fees for overhaul at various rates. Residents renovated house pay a smaller amount for it.

The rate laid a systematic approach to the estimation of parameters – design features and the age of the building, the quality of the roof materials, window and door openings. This experience can be of practical interest for the management of municipal services Russia [11]

Table 3. Housing and communal services actually paid by the population, mln rubles [5]

Regions	2007	2009	2011	2012	2013	2014
Russian Federation	722 524,5	1 063 282,7	1 466 772,3	1 533 237,1	1 677 213,5	1 820 164,1
Central Federal District	225 865,1	332 734,8	460 690,7	482 180,6	530 352,9	577 388,9
Moscow	72 965,9	103 026,8	132 517,6	138 448,7	154 140,6	167 259,9
Northwestern Federal District	85 123,5	123 479,9	179 418,5	189 092,4	196 528,9	220 809,7
Saint Petersburg	31 191,4	45 720,5	68 858,1	74 547,4	74 811,1	92 816,3
Southern Federal District (till	74 415,9					

DOI: 10.1051/e3sconf/20160601004

2009)						
Southern Federal District (from 2010)		81 770,9	112 098,7	121 029,4	130 694,9	146 278,
North Caucasian Federal District		30 506,3	43 177,4	46 235,3	54 993,2	57 614,5
Volga Federal District	136 259,2	205 541,5	276 353,3	287 380,4	319 062,5	344 207,
Ural Federal District	66 576,1	93 040,3	131 940,7	135 518,9	150 343,8	162 250,
Kurgan region	3 077,6	4 792,7	6 624,7	6 922,9	7 650,8	8 162,3
Sverdlovsk region	21 249,3	31 106,4	46 241,	48 157,5	54 153,8	59 832,3
Tyumen region	26 028,	35 992,5	45 022,8	44 992,5	49 281,	52 165,
Khanty-Mansi Autonomous District	14 626,9	19 762,9	25 417,5	25 445,4	27 349,3	28 520,3
Yamalo-Nenets Autonomous District	6 025,5	8 482,1	9 665,4	9 371,5	9 361,4	9 590,8
Tyumen region (without AD)			9 939,9	10 175,5		14 053,8
Tyumen region (in part)		7 747,5	9 939,9	10 175,5	12 570,3	
Chelyabinsk region	16 221,2	21 148,7	34 052,2	35 446,	39 258,3	42 090,4
Siberian Federal District	88 005,9	128 842,7	171 905,	178 239,	192 129,9	204 469,7
Far Eastern Federal District	46 278,7	67 366,3	91 187,9	93 560,9	103 107,5	107 146,2

The table shows the amount of payments for housing and communal services taking into account the payment of debt from previous years in the dynamics of the odd years (2007 to 2015).

3. Late payment or refusal to pay contributions for the overhaul.

This problem is related to the relationship of power and property owners. Over the past six months by region swept by a wave of discontent: contributions to overhaul turned the problem of mass non-payment and lawsuits about the validity of the board [12].

Crisis in 2015 reduces the solvency of the population. The risk of unemployment and little income increased the number of "non-payers".

Income distribution system has the risk of misuse of funds, conspiracy irresponsible contractors and "lazy" system placing temporarily free funds.

Temporarily free funds should work for growth. At present time there is not even the interest on the accounts of FO and condominiums even in the amount of a key Bank of Russia. This situation shows the pseudo-effective and can lead to enormous loss of inflation. It is advisable to place a temporary surplus in low-risk financial instruments (regional bank deposits). This will increase the lending capacity of regional commercial banks and to increase employment and provide economic development of regions.

Transparency of information on the use of funds and contracting will provide free access to homeowners to electronic resources of the Regional Fund for the Promotion overhaul of the common property.

The question of allocation of temporary free funds remains open. FO accumulate funds in bank accounts (FO Sverdlovsk region began to accumulate funds for the Agricultural Bank account). What lies behind this?

There is a latent additional funding government programs and ensuring the liquidity of the largest banks with state participation, which are financial "safety cushion" of local authorities.

Innovation 2. Licensing management companies (MC).

The introduction of licensing is intended for HCS market clearing of irresponsible management and destruction of the "gray" market. Already in May 2015 monitoring revealed twenty-three "empty" company in Yekaterinburg [13].

Experience has shown that a number of countries (USA, Germany, France, Finland, Poland, Hungary), the licensing administration increased the quality of municipal management. On the housing market only the best companies. In addition, the increased specialization and increased competition between the same specialization of organizations [14]

Browse companies control the licensing procedure in accordance with the law has highlighted the problem of quality control management in the housing sector.

These problems are internal (from the perspective of business processes management companies themselves) and external (from the point of view of the impact of external factors) character.

On the one hand, MC which is in charge of a lot of illiquid housing facilities can throw off this burden from the shoulders and give up to obtain a license. The dilapidated legacy" take yourself municipal budgets. On the other hand, it is easy to obtain a license scheme will wish to establish fraudsters management companies to raise money from the population. Withdrawal of a license – a slow process, during which time the company will provide fraudulent financial "safety cushion".

2014–2015 marked in Yekaterinburg "communal wars" between rival management companies. These events are taking place against the backdrop of awakening civic consciousness owners [15].

An element of the state policy in relation to the MK is the application of penalties for various violations: the concealment of information, fraud in the calculation of rent, violation of the rules of detention houses. These penalties were tightened in 2015. However, the MK is commercial organization and the main purpose is profit, which comes from the "pocket" residents. Accordingly, the penalties – is the hidden fees, included in "maintenance of housing." In addition, the cost of utilities will be included costs for licensing non-core expenses.

Table 4 shows the amount of fines paid to the management companies of Yekaterinburg in 2014 for various violations.

Table 4. Amounts of the fines paid by the housing management companies of Yekaterinburg in 2013 [10]

№	Management company	Amount of fine (Thousand Roubles)
1	MC Standard / ЗАО «УК Стандарт»	1 935,5
2	MC HCS Octyabrsky district / ООО «УК ЖКХ Октябрьского района»	818,5

3	MC Verkh-Isetskaya / ЗАО «УК Верх-Исетская»	795,6
4	MC REMP Zheleznodorozhny district / ЗАО «УК РЭМП Железнодорожного района»	704,9
5	Fund Radomir / ООО «Фонд Радомир»	510,1
6	HMC Radomir-Invest / ООО УЖК «Радомир – Инвест»	411,0
7	HMC Ordzhonikidzevskaya / ЗАО «Орджоникидзевская УЖК»	381,0
8	MC Nizhneisetskaya / ООО «УК Нижнеисетская»	358,0
9	MC REMP UZHSK / «УК РЭМП УЖСК»	343,0
10	MC REMP Elmash / «УК РЭМП – Эльмаш»	317,0
11	MC HCS Ordzhonikidzevsky district / ООО «УК ЖКХ Орджоникидзевского района»	282,0
12	HMC Ural-ST / ЗАО «УЖК Урал-СТ»	213,0
13	MC Chkalovskaya / «УК Чкаловская»	199,0

New measures of state influence on the management companies may indirectly contribute to the destruction of competition and redistribution of the housing market in favor of the companies affiliated with the authorities. In practice, this scheme can be implemented as follows: the management company is deprived of the license as prescribed by the state supervisory authority for the poor service 15% of the total floor area of all buildings served. Then the "objectionable" economic counterparties are removed from the playing field and come into play asset managers, "nurtured" by the authorities.

In the area of control over the activities of management companies are invited to apply the experience of European countries. For example, in Sweden, Italy, the UK is extremely well developed "house" movement with a horizontal structure of relations. Association homeowners monitor tariffs and quality of service. They protect the rights of owners in court.

International Alliance of Inhabitants (IAI) helps citizens in the implementation of social rights. Housing Swedish Court prevents corruption. The tariffs of housing and communal services are transparent. Since 2006, the European people's universities are functioning housing, where the population is trained active government technology [16].

An effective method of improving the financial performance of management companies can act as a requirement to increase the authorized capital, which will not only guarantee the satisfaction of creditors' claims, but will also get an idea of the scale of operations and financial condition of the management company.

As a method of minimizing the risks and at the same time one of the most effective tools of financial stabilization of the management companies can recommend the creation of reserve capital.

As part of the retained earnings capital will guarantee the obligations liquid unforeseen expenses and possible losses. Provisions allow management companies to reserves will increase the stock of financial strength and to reduce operation risks in terms of survival.

Innovation 3. On January 1, 2016 changed the treatment system of waste management, which became known as "utilities" [17]. Regional operators collect and recycle waste.

However, terminological innovation has led to a lack of clear understanding of the definition of "municipal solid waste" left unresolved questions about the guarantees of quality treatment services and the waste of the area of responsibility for servicing them. For example, "without a master" stayed container sites for waste collection.

Problems of waste management, resulting in the "management" of man, put off no longer: it is a question of security of human living in Russian cities. We offer the best experience to appeal to the European Union countries on the collection and disposal system. The most effective system of waste management, improves the environment, there is in Sweden. Burial subjected to only 1% of the waste, the rest falls into recycling [18].

The authors believe that it is necessary to conduct a thorough classification of morphological composition of municipal waste and to allocate clear lines of responsibility of municipal services managers. Urban management is a holistic system. It is impossible to address issues of financial or technological support apart from psycho-physical laws of interaction between people in the society. Urban management in Russia is currently showing contradictory results.

Appeal to the experience of effective functioning of the management models do not allow the "roll down" to "false" understanding urbanization as a post-response system of urban management problems. It is today we need the expert judgment of the urban space with its social contrasts and environmental constraints.

References

1. Unified interdepartmental information and statistical system, http://www.fedstat.ru
2. E. Kondyukova, O. Nikitina et al., Construction economy, **4**, 46 (2015)
3. E. Kondyukova, O. Nikitina et al., Construction economy, **2**, 14 (2016)
4. Federal Law of 29.12.2014 No. 458-FZ (ed. of 29.12.2015) "On Amendments to the Federal Law" On Production and Consumption Waste", Separate Legislative Acts of the Russian Federation and the Annulment of Certain Legislative Acts (Provisions of Legislative acts) of the Russian Federation" (rev. and ext., joined. in force from 01.01.2016), http://base.consultant.ru/cons/cgi/online.cgi?req=doc;base=LAW;n=191899;div=LAW;dst=100004,0;rnd=203280.9831487736089928
5. The Federal State Statistics Service, http://www.gks.ru/
6. Regional Fund to facilitate capital repair of common property in apartment buildings of the Sverdlovsk region, http://www.fkr66.ru
7. Federal Payment Standards of housing and communal services in 2013–2015, Resolution of the Russian Government dated February 21, 2013 No. 146 (repealed), http://base.consultant.ru/cons/cgi/online.cgi?req=doc;base=LAW;n=142431;div=LAW;dst=100003,-1;rnd=203280.10032736270715814
8. On establishing in 2014 the minimum fee for the overhaul of the common property in apartment buildings, located on the territory of Sverdlovsk region, Resolution of the Sverdlovsk Region Government No. 1625-PP dated 27.12.2013, http://base.consultant.ru/regbase/cgi/online.cgi?req=doc;base=RLAW071;n=128829;div=LAW;dst=0,0;rnd=203280.8920819437326677
9. On the establishment the Sverdlovsk Region minimum fee for the overhaul of the common property in apartment buildings on 2015–2017, Resolution of the Sverdlovsk Region Government No. 833-PP dated 10.01.2014, http://base.consultant.ru/regbase/cgi/online.cgi?req=doc;base=RLAW071;n=158401;div=LAW;dst=100003,0;rnd=203280.3387107734360899
10. Sverdlovsk Government Web-portal, http://www.uralinform.ru
11. Housing and communal services tariffs in the Russian Federation in 2000–2010, Analytic report, Energy and Finance Institute, http://genproc.gov.ru/upload/iblock/a36/bhmyrs_epjuxlazyj%2014.01.11.doc
12. Web-Portal of federal arbitration courts of the Russian Federation, http://www.arbitr.ru
13. State information system of housing and communal services, http://dom.gosuslugi.ru
14. G. Makarov, Economy and Law, **6** (2011)
15. The web-portal of the Russian Federation Ministry of Construction, http://www.minstroyrf.ru
16. The official website for European Social Forum, http://www.fse-esf.org

DOI: 10.1051/e3sconf/20160601004

17. The Russian Federation Housing Code of 29.12.2004 No. 188-FZ, http://base.consultant.ru/cons/cgi/online.cgi?req=doc;base=LAW;n=193170;div=LAW;dst=100003,1;rnd=203280.4900346251295875
18. Alternative energy technologies web-site, http://www.alterenergy.info

Impact of Anthropogenic Factor on Urboecological Space Development

Tamara Kuprina[1a], and *Lyubov* Sergeeva[2]

[1]Ural Federal University, International Economics Department, 620002, Yekaterinburg, Russia
[2]Ural Federal University, Fundamental Education Department, 620002, Yekaterinburg, Russia

Abstract. The article discusses the issues of the impact of the anthropogenic factor on urboecological space development. The issues are considered taking into account retrospective theoretical data to show the process of Anthropoecology development as a new branch of sociological science. At present the noosphere acquires features of anthropoecosystems having a number of parameters from the endogenous and exogenous point of view. Anthropoecology has special socio-cultural significance as considers the interaction of all actors of international space. There introduced the new branch Ecopsycology as the outer world is the reflection of the inner human world. There is a definition of the sustainability of ecological system. In the practical part of the article there is an example of academic mobility as the basis of the human potential with possible transfer into the human capital supporting by survey data. In conclusion there are recommendations on management and adaptation of the anthropogenic factor (a kind of biogenesis) in modern urboecological space.

Today's global challenges raise a number of major transformations in all spheres of the society life. However, it should be noted that, in speaking of the system of comfortable conditions creation for living in urboecological space, experts generally have in mind the technical, political, economic, legal aspects, often underestimating the humanitarian component: the human factor. This factor is important, first of all, because all ecological problems are generated by the people themselves.

At present, there is a separate branch of knowledge: Human Ecology (Anthropoecology). And, if earlier it concerned mainly the physiological state, now there is an interdisciplinary approach that includes social ecology, urban ecology, culture ecology and others.

The proposed concept includes not only the traditional consideration of ecology of human relations and society with nature, but also the relationships between the structural units of the society. In Anthropoecology the object of research is social processes in relation to the state of the subject (person, group of people, including intercultural community), which appear in different areas of life.

The area is closely related to Urboecology, studying not only the features of the ecological space of the city and urbanization, but also the influence of anthropogenic

[a] Corresponding author: tvkuprina@mail.ru

factors on the development of the urban space. Thus, it is necessary to study the human factor, constantly moving from one ecological niche to another (housing, work and study places, transportation, recreation, etc.). Moreover, with increasing urbanization and space alienation processes people must have the complex adaptive mechanisms (physiological, psychological, social) that are not limitless, but they can be developed and improved. In particular, it is necessary to take into account the cultural and historical space of the city, including the social and spiritual environment, considering historical processes, cultural traditions prevailing in the country and region specified.

In today's Urboecology one must take into account the demographic and migratory factors. Transnational migration flows involve not only the labor and professional migration, but academic migration as well, the number of which increases with each passing year. The high degree of academic mobility requires a well-developed infrastructure (hostels, health insurance, etc.), availability of financing sources (scholarships, travel grants, etc.), high language skills of students and lecturers enable to teach and study abroad as well as the psychological readiness of academic migrants to adapt to the new conditions.

Consequently, practical activities should be aimed at optimizing the interactions between all the actors involved: analysis of interacting social structures in different socio-cultural contexts, their adequate assessment and interaction with different gender, ethnic, age and other groups under the urbanized space. It is necessary to take into account the experience and development features of other cultural communities.

Some authors distinguish Anthropoecology as a separate field and as an integrated discipline that combines elements of ecology and anthropology, and studies the human activity, interacting with nature, society and culture. There are factors determining the quality of life, individual values and the problem of their improvement. Moreover, this approach to the solution of educational problems raises from the fact that human is the highest value and measure of effectiveness of civilization as a way to organize the human lives and communities (a prosperous society) [1]

On the other hand, Anthropoecology researches the relationship laws of the biosphere and human anthrosystem, their groups and individuals, the impact of natural and social space on the individual and their groups, human populations, including ethnic groups [2].

Thus, Anthropoecology is an integrated discipline, where all social, economic, natural and cultural factors are considered as equivalent components of human lives, providing different levels of their needs.

Theoretical concepts are based on different approaches to the interpretation of social interactions in ecological space. However, yet Aristotle, F. Akvinsky, N. Machiavelli clearly traced parallels between natural and social development of the society.

Auguste Comte drew attention to the labor division and solidarity. E. Durkheim pointed to the cultural artifacts and linked them with the collective representations. M. Weber considered mutually oriented or social actions of people as a unified factor.

In XIX century, due to H. Spencer, there appeared a new social and biological direction, interpreting living beings from a social point of view. At the same time there formed Ecology as a discipline (E. Haeckel, 1866), connecting biological and sociological concepts.

R. Park and E. Burgess (Chicago Social School, 1921) founded the direction "Human Ecology", which later received other names: Anthropoecology and Environmentology. Its purpose is to define the processes of interaction between the environment and human communities as well as their impact on human activity.

A. Tensli (1935) introduced the term "ecosystem" (Greek oikos – home, settlement, and systema – combination, association), which is understood as a set of co-inhabiting organisms and the conditions of their existence, being in a logical relationship with each other and forming a system of interdependent biotic and abiotic processes and phenomena [3].

A. Tensli emphasized that ecosystems were the main natural units comprising both a complex of organisms and physical factors that made up biomes' space, i.e. habitat factors.

In the Russian science V. Sukachev (1940) coined the term "biogenesis" as a homogeneous area of the earth surface with a defined composition of living (biogenesis) and inert (surface layer of the atmosphere, solar energy, soil, etc.) components, integrated by metabolism and energy exchange in a single natural complex. Moreover, ecosystems can have arbitrary boundaries (from a drop of water to the biosphere as a whole), while the biogenesis always occupies a certain territory [3].

As noted by E. Kogay, ecosystems strive for balance between the number of organisms and habitat capacity (the ability of habitat to allow the ecosystem to exist without affecting its constituent components). This balance is not static; it is flexible, changeable within certain limits defined by the terms of the maintenance of the basic parameters of the ecosystem functioning. Going beyond these limits (often by human intervention) can lead to destructive processes in the system or changing its fundamental components [4].

V.I. Vernadsky and then his follower A.E. Fersman point out that with the development of industrial activity a human becomes a major geological factor of all changes on the planet. Thus, the biosphere transforms into a qualitatively new state, indicated by V.I. Vernadsky as the "noosphere". Moreover, technical and social backgrounds of the noosphere appear long before the human awareness of the need to shift to the noosphere.

In this case, technical and social prerequisites for the noosphere are: a sufficiently high level of technological development, energy security and scientific knowledge; overcoming the economic, political and racial separateness of people with absolute exclusion of any military conflicts from the interrelations between the peoples. ... An important social condition of the noosphere is a need of transition of all humanity to a higher degree of social integration. Humanity can only survive as a single entity [5].

One of the transitions of the biosphere into the noosphere is the process of urbanization. The city comprehensively creates many aspects of human life. In assessing the degree of the ecological comfort of the city one must take into account, in particular, the components of the citizens' life, as the level of social welfare (family budgets, housing, services, education, health, quality of medical care and social security, etc.), the degree of ecological safety and legal protection, employment and satisfaction with work (scope of employment, relationships at work, transport and pedestrian accessibility of the work place, etc.), the conditions for a good rest and recuperation, the completeness degree of information support and the existence of conditions for the continuity of cultural traditions etc. Thus, the macroenvironment of the city consists of numerous microenvironments that impact the well-being of people [6].

In modern conditions there is a new kind of noosphere: anthropoecosystem. From B.B. Prokhorov's point of view, anthropoecosystem is a spatial subdivision of the human environment, in all its parts having the similarity of natural, social, economic, industrial, ecological and hygienic, cultural and domestic, informational conditions of human life, which form the worldview and environmental awareness, health level, demographic behavior, physical appearance, labor skills, lifestyle, customs and traditions, choice of religion, professional preferences, etc. Anthropoecosystem is an ecosystem in which human activity takes place [7].

Anthropoecosystem has a number of characteristics:

1. Internal homogeneity and external heterogeneity with other similar systems;
2. Takes a certain space;

3. Self-development is gradual over long periods of time;
4. Rapid change gives rise to new systems and often undesirable phenomena for the human society.

Anthropoecosystem includes a variety of elements such as nature, economy, population, socio-economic conditions. Moreover, all the elements are interconnected and changes in both individual elements and their combinations impact on the human behavior, whose community is a central element of the system. Thus, there are changes in environmental awareness, demographic and professional behavior, health, culture and education levels.

In this context, one can talk about socialization, i.e. specific adaptation of a human or group of people to the social environment. Consequently, there is a need to ensure the conditions meeting the needs and interests of the people and contributing to the achievement of important vital goals.

L.K. Kruglova allocates a separate direction, i.e. socio-cultural Anthropoecolology, whose subject is laws of the structure, functioning and development of the ecosystem, elements of which are human, nature, society and culture [8].

Of particular relevance to human is such an element of cultural microenvironment, as home of a wide socio-cultural sense. Home means a family, its spiritual atmosphere, life style, objects in the house, and, finally, the very image of home, that is a kind of unity. It is at home where a human has an opportunity to put into practice all the principles of interaction with the socio-cultural environment, all the ways of overcoming its negative states [8].

The newest trend is Ecopsychology, which tries to combine the transformation of the individual and society. Ecopsychologist Jean-Pierre Le Dunff notes that according to the tenets of the science, the outside world is a reflection of the inner world of humans [9].

In general, the subject of research of ecopsychology is ecological consciousness as a set of ideas about the relation to nature, strategies and technologies of interaction with it on socio-genetic, ontogenetic and functional levels.

In this regard, one can speak of a certain ecosystem sustainability contributing to the sustainable development of the society. For example, E. Tsukanov offers the following definition of ecosystem sustainability: the sustainability of the ecosystem is an ability to withstand environmental fluctuations, while retaining its structural and functional integrity [10].

However, sustainability can constantly be disturbed by external factors. One of the most difficult aspects of modernity is the problem of international migration, as demonstrated by the migration crisis in Europe which has received millions of socially non-adapted representatives of many cultures, dramatically different from the European one. The protective reaction of migrants can be considered as the formation of ethnic enclaves in the host areas, preserving the behavior standards and traditions of national cultures.

At the same time, there are other "soft" kinds of migratory flows, in particular, the academic migration. The advanced format of the flow includes not only students but also post-graduate and doctoral students, students of preparatory and language courses, seminars, internships, i.e. it is a migratory flow caused by getting an education in the broadest sense [11]. In the Russian migration flow academic migration takes the third place after the labor migration and resettlement of compatriots.

Researchers of academic migration show that every year there is an increasing competition between countries for academic migrants as foreign students are both a norm of international cooperation and "soft power" in the context of the geo-cultural space, and future "quality labor migrants" (young, embedded in the socio-cultural context). The increase share of academic migration in the total migration flow provides greater manageability of the immigration process; consequently, the academic migration can be considered an important component in the security concept of any country [12].

However, there are several problems associated with adaptation of academic migrants in the new intercultural space.

First of all, experts argue that our time is the era of customized environment. The younger generation "customizes" the world, making it more convenient fro them. Students should be able to "adjust" to some extent, the learning process, so that each of them has the most convenient way to gain knowledge [13]. It is necessary to add that they customize not only training, but also urbanized space.

Therefore, the Russian education system faces an issue of life and study optimization of foreign students, who are going through a difficult process of adaptation to new conditions of life.

There is no doubt that in order to attract foreigners to study the quality of the living conditions for students is of great importance, as each country has its own peculiarities of everyday life and foreign students have to accept different norms and rules of behavior, communication, keeping household. The important problem for foreign students is the language barrier. In order to improve the learning process of foreign students it is necessary to develop their professional and linguistic competence [14].

To determine the specific problems of academic migration adaptation in the urboecological environment, we conducted a pilot survey among foreign students at first-year full-time training. The sample consists of 40 people, arrived at the Ural Federal University (UrFU) in 2015. We use the copyrighted methodology by I.A. Sholokhov, aimed at identifying the difficulties in communicating in a foreign language, the presence of the motivation to study and determine the total adaptation of students for life in the new environment and identify the students' loyalty to another cultural society. The survey involved students from 20 countries, 16 far abroad countries (Kenya, Zambia, Mongolia, Mexico, China, Pakistan, Egypt) 4 the CIS countries (Kyrgyzstan, Tajikistan, Kazakhstan, Uzbekistan).

The results show that 37 out of 40 have problems in intercultural communication, associated with lack of knowledge of a foreign language or a reluctance to join the cultural environment, to learn and adapt to new conditions. Students are not used to the new conditions of life, and the adaptation is difficult. Only 3 students out of the total sample identify themselves with the student environment and strive to fully adapt to it, to learn, to immerse in the new language environment, they have a positive motivation for learning activities.

Answers of half of the respondents indicate inadequate assessment of their position in the cultural environment and the poor adaptation to changes in living conditions. 10 respondents have no active socio-cultural position. 8 out of 40 (20%) are loyal to the cultural environment in which they are located, respectively, they can successfully adapt to the new conditions of life. 2 respondents have not formed a clear position yet, and there is no particular motivation to educational activity, it is likely that students can stop learning.

There are some issues related to the urboecological adaptation.

1. "Settlements of foreign students should be focused, but not dispersed across the city". Students' opinions are as follows: "Yes" – 25 respondents, 10 – "No" and 5 – difficult choice. It also shows that the majority of the respondents prefer to live in the same place with people of their own ethnicity, close to the university, as their navigation in the city is difficult, it also means a low adaptability level to the life conditions in another country.

2. "Development of services for international students is their own concern, rather than the local authorities." The students respond as follows: 10 people agree with the statement, 21 people do not agree with the expression and 9 people found it difficult to answer. More than half of respondents believe that the regional authorities should take care of the development of services for international students. This issue is a control point for

a series of issues about the student's loyalty to the community and other cultures and the answer "do not agree" indicates the inadequate assessment of their position in the cultural environment and bad adaption to changes in living conditions.

3."Would you like to have more national dishes for the foreign students in students' canteens?" 34 respondents answer "Yes", 3 – "No" and 3 – difficult choice, it shows that students miss home, their national cuisine, and that they are not adapted to local nutritional habits.

4."Educational institutions should provide special services for international students." The majority of respondents (35) answer "agree", 2 – "disagree" and 3 – difficult choice.

5. "With respect to foreign students social norms need to be more flexible" 33 respondents reacted to the positive response, only 3 people did not agree with the opinion and 4 difficult to answer.

Two last issues determine the overall adaptation of students for life in a new environment and positive answers "agree" of respondents suggest that adaptation is difficult and students are not used to the new space conditions.

Thus, the answers to the issues of urboecological character show both the weak adaptation of foreign students, and underdeveloped urbanized structure of the city, which should support the invited students, who could later become professionals in the area.

The presented pilot research compares favorably with most other ones as they are mainly based on ordinary questionnaire surveys, limited in number and aspects discussed. As far as the research presented in this article, it is based on a complex professional psychological tests that validity has been proved. Despite the limited sample (40 respondents) which is due to the beginning of the research, it can be a good example proving some data received by other experts.

So, according to their survey, foreign students admitted the following problems: climatic (26%), hostel conditions (21%), communication in the Russian language (18%), new life style (14%), natives' attitude (11%), relatives' absence (5%) and Russian cuisine (3%) [15].

From our point of view, the extended research planning by the authors of the article can help better understanding of the problem and formulating new aspects of its solution.

In foreign universities for successful adaptation of students there are various mechanisms of socialization and adaptation. Typically, the host country students are engaged in the process by organizing the so-called clubs. If we consider the European experience, it is necessary to mention Buddy Building Club (BBC), the international volunteer club, which deals with the adaptation of foreign students on a voluntary basis [16].

Recently the similar project has been introduced in the Ural Federal University. "Buddy UrFU" is the foreign student support program aimed at creating a favorable environment for learning and socialization through the involvement of volunteers among students to meet and support foreigners on their first most difficult days of stay in the new country and help in solving a variety of domestic issues. Between August and December 2015 "Buddy UrFU" helped 700 foreign students.

The Ural Federal University has also set up its own associations to support foreign students, for example, Student United Nations Organization, the students' organization, formed in February 2012, in order to provide comprehensive assistance to foreign students to adapt to student's life at the university. Its main task is in continuous contact with the foreign students during the whole period of training (through students' fraternities that are created on the principle of territorial or national identity). The fraternities help students to adapt, monitor educational progress and help in solving problems. At the moment, there are 12 students' fraternities groups, uniting more than 70% of foreign students.

Taking into account the theoretical principles mentioned above, and the practical results of our pilot survey, we can give the following recommendations:

1. Considering complex migration and demographic situation, academic migration should be seen as a possible source of human potential, followed by the formation of high-quality human capital in the host urboecological space.
2. Assessing urboecological comfort of the city for receiving academic migration it should be taken into account the comfort of living conditions and their destination to socially significant objects.
3. Special attention should be paid to health and timely quality health care, as it is due to the difficulties of foreign students' adaptation to the climatic conditions. On the other hand, there is a need for prevention of Russian students' morbidity because foreign students can be carriers of unusual diseases for this urboecological space.
4. From the intercultural point of view, there must be conditions established for adaptation to new cultural traditions and norms of behavior, as well as the prevention of ethnic conflicts, taking into account the compact settlement of the fraternities' type.
5. The above-mentioned students' organizations will certainly help their fellows to adapt to the host urboecological space. However, from our point of view, it is not enough. Support of academic migration, i.e. the quality human potential, and subsequently the human capital what is socially and economically important in terms of demographic problems, should be dealt by official city agencies.

Thus, we can conclude that there are changes not only in natural and technological landscape in the urboecological space, but also in the demographic situation due to international migration. In this case, one can consider a certain biogenesis, anthropogenic factor, creating its own special urboecological space. Consequently, the human factor is becoming increasingly important defining features of the designing and maintaining urboecological systems.

References

1. S.V. Krivikh, A.A. Makarenya, *Pedagogical Anthropology* (SPB, 2003)
2. N.A. Agadzhanyan, *Human Ecology* (1997)
3. M.S. Gilyarov, *Biological encyclopedic dictionary* (M, 1986)
4. E.A. Kogay, *Doctrine of biosphere and biosphere science class*, SHK, **3**, 92–105 (1999)
5. E.V. Girusov, *Ecology and Economics of wildlife management.* Available: http://nashaucheba.ru/v41859//
6. *Urboecology and Anthropoecology*. Available: http://ecology/00170400_0.html
7. B.B. Prokhorov, *Human Ecology* (2010)
8. L.K. Kruglova, *Socio-cultural Anthropoecology as methodological basis for personality typology*. Available: http://www.pandia.ru/text/78/461/6808.php
9. A. Grigoryeva, *How to overcome ecological fatigue*, Ps, **3**, 122–124 (2016)
10. E. Tsukanov, *Information society and Homo Informaticus*. Available: http://www.bibliofond.ru/view.aspx?id=116652
11. O.D. Vikhovanets, *Academic migration as a part of migration policy in Russia*. Available: http://www.fms.gov.ru/about/science/science_session/forth/vih.pdf
12. E.S. Krasinets, *International migration of Russian population in conditions of market transfer* (2012)
13. D. Koats, *Generational learning styles* (2011)
14. L.V. Sergeeva, *Demographic processes impact on the development of academic migration in Russia* (Yekaterinburg, 2015)
15. A.D. Gladush, G.N. Trofimova, V.M. Filippov, *Socio-cultural adaptation of foreign citizens to the conditions of studying and living in Russia* (2008)

16. E.N. Shpet, *Mechanisms of social adaptation of students-migrants learning under academic mobility programs*, SI TPU, **11**, 170–172

E3S Web of Conferences **6**, 01006 (2016) DOI: 10.1051/e3sconf/20160601006
ICSC 2016

Municipal solid waste management in a new legislation: comprehensive approach

Maria Berezyuk[1], *Alena* Rumyantseva[1], and *Alena* Lapina[1a]

[1]Department of Environmental Economics, Ural Federal University, 620002, Yekaterinburg, Russia

Abstract. The problem of the municipal solid waste (MSW) formation and recycling has been very important for many decades in the Russian Federation. The sustainable development of the Russian Federation subjects and municipalities, their evolution to the status of "smart cities" is not possible without solving this problem. The current situation in the area of SHW treatment leads to the dangerous environmental pollution, improper use of the natural resources, and the significant economic damage and poses a threat to the health of the present and future generations of the country. The authors examine the issues of implementation of the changes in the field of municipal solid waste management legislation. The problems of the comprehensive waste management are also discussed on the example of the Sverdlovsk region (Sverdlovskaya Oblast).

1 Municipal solid waste formation as a kind of negative impact on the environment

The sustainable development of the modern society is a comprehensive result of the whole number of components. The environmental component becomes more and more critical with every year. For almost all subjects of the Russian Federation one of the main task in the field of environment protection is solving the problem of neutralization and reclamation of the waste, among which the biggest problem is the municipal solid waste generated in the residential area, due to the complex morphological structure and uneven distribution of the formation sources. The amount of the industrial and consumption waste generated in the Russian Federation is 5,168.3 mln t., the share of the used and neutralized waste was 45.6%, i.e. 2,357.2 mln t. in 2014. For comparison, the amount of the generated solid municipal waste was equal to 56.68 mln t. and the share of the used and neutralized waste was just 7.5% or 4.26 mln t. The dynamics of the formation and use of the industrial and consumption waste and municipal solid waste in the Russian federation is shown on Figure 1, 2 [1].

[a] Corresponding author: alenalapi@yandex.ru

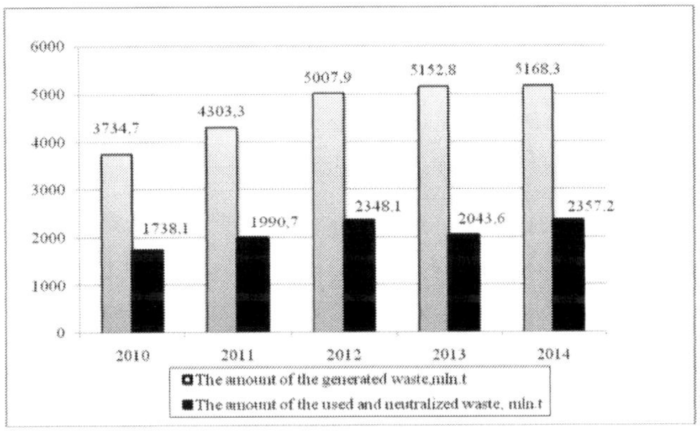

Fig. 1. The dynamics of the industrial and consumption waste generation and use in the Russian Federation.

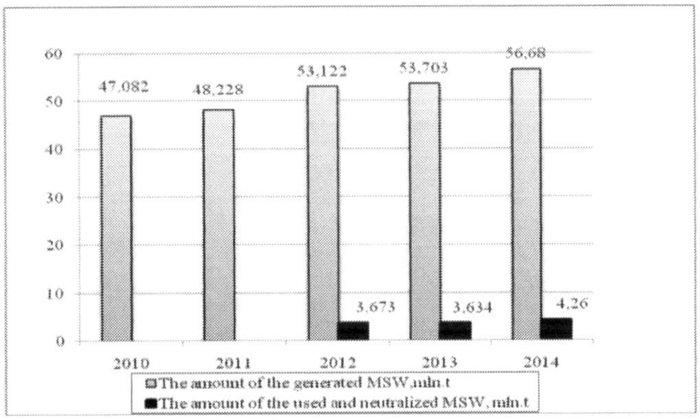

Fig. 2. The dynamics of the MSW generation and use in the Russian Federation.

The sanitary cleaning of the inhabited localities is one of the most complex and important challenges faced by the local municipal autonomous bodies at the territory of the Sverdlovsk region. At the same time the amount of the MSW is increasing and the territorial opportunities for its disposal are decreasing.

The data on the amount of the MSW generation in Sverdlovsk region in 2014 are presented in Table 1 [2].

Table 1. The amount of the MSW generation in Sverdlovsk region in 2014.

Index	**Region in total**	**The administrative district of the Sverdlovsk region**					**MF «Ekaterinburg city»**
		Eastern	**South**	**Northern**	**Mining & Metallur-gical**	**Western**	
Municipal waste generation amount, ths. t.	1775.9	310.13	269.2	175.97	400.84	189.9	568.
including the solid municipal waste	1 540.16	89.38	254.89	175.88	400.7	189.86	117.87
actual municipal solid waste generation per one resident of the district, t./1 resident	0.36	0.18	0.41	0.37	0.58	0.32	0.083

For 2014 4.013 business entities reported on municipal waste treatment using the Technical report form (3.973 entities in 2013), which accounts for 96.1% of the total amount of the reported entities [2].

The comparative analysis of the municipal waste treatment during 2013–2014 is not carried out due to the Federal waste classification catalogue (FWCC-2014) entering into force, in which the group of the municipal waste does not include the types of waste that earlier belonged to it. The data regarding the generation, use, neutralization, placement, and availability of the municipal waste by the end of 2014 are presented in Table 2.

Table 2. The data regarding the generation, use, neutralization, placement, and availability of the municipal waste by the end of 2014 [2].

Index	ths. t.	%
Generated, in total	**1 775.9**	**100.0**
among them: municipal solid waste	1 540.2	86.1
municipal liquid waste	235.7	13.9
Used and neutralized by the business entities, in total	**200.5**	**100.0**
among them: municipal solid waste	198.9	99.2
municipal liquid waste	1.6	0.8
Placed, considering the temporarily stored wasteв, in total	**1 523.4**	**100.0**
among them: municipal solid waste	1 289.7	84.6
municipal liquid waste	233.7	15.3
Waste availability by the end of the year	**59 768.5**	**100.0**
The amount of business entities reported on the municipal waste	4013 out of 4175	

In the region there is an unfavorable situation with the organization and operation of the municipal waste dumps.

Out of 429 facilities accumulating municipal waste 284 (66.2%) have no defined business entities operating the accumulation facilities; 229 (53.4%) facilities have no hydrogeological conclusion; 262 (61.1%) facilities have no formal documents for the land tenure, landownership. One of the basic requirements for the dumps operating – the existence of the project – is fulfilled only by 55 waste disposal facilities (12.8%).

2 Legislation changes in the field of MSW treatment

Over the past two years the environmental legislation of the RF has undergone major and dramatic changes. In December of 2014 the Federal Law No. 458 "On the Amendments of the Federal Law "On production and Consumption Waste", certain legislative acts of the Russian federation and the annulment of certain legislative acts (provisions of the legislative acts) of the Russian Federation" (here in after Federal law No. 458) was adopted. This document dramatically changes the system of the waste treatment affecting the fundamental concepts and principles of the environmental legislation. The changes affected practically all aspects of the waste treatment, opening the subjects regarding the ownershift of the waste, the licensing of the waste treatment activities, professional training of the persons admitted to the waste treatment, and many others. In addition, the terminology used in the field of waste treatment was also changed. The innovations affected the most often generated waste – municipal solid waste [3].

Municipal Solid Waste (MSW) is the waste generated in the living quarters during the consumption by the individual persons as well as the goods that lost their consumer properties during the process of their use by the individual persons in order to satisfy their personal and domestic demands. The waste generated in the process of the legal entities activity, independent businessmen activity and similar by the structure to those generated in the living quarters during the consumption by the individual persons (Art. 1 of the Federal Law No. 89).

On January 1st, 2016 the Federal Law No. 89 was supplemented by a whole chapter – chapter V. 1 **"Regulation of the activities in the field of the municipal solid waste treatment"**.

Thus, according to the changes, the **regional operators** will be appointed for the activities in accumulation, transportation, treatment, recycling, neutralizing, and disposal of the MSW at the territories of the Russian Federation subjects.

Regional operator for the MSW treatment is a legal entity, which is required to conclude a contract for providing services for the treatment of MSW with an owner of the MSW, which are generated and accumulated in the regional operator's zone of activity (Art. 1 of the Federal Law No. 89 of 24.06.1998 "On the Production and Consumption Waste").

The regional operators should carry out their activity in accordance with the regional program in the field of the waste treatment and territorial scheme of the waste treatment, the development and approval of which will be within the competence of the Russian Federation from January 1st, 2016.

In 2016, besides the fundamental changes in the environment legislation, the dramatic changes in the field of housing and communal services legislation are planned. Thus, instead the service *"removal of the waste"* a new municipal service is introduced, which will be called *"municipal solid waste treatment"*. The introduction of the new service will entail the changes of the procedure of calculating the amount of payment for collection and disposal of the MSW. Currently the amount the amount of payment for the same service is determined in anticipation of 1 m^2 of the total area occupied by the consumer premises. According to the changes in the Federal Law No. 89 the fee will be calculated on the bases of MSW accumulating norms per 1 person [4].

For the regional operator's service of treating the MSW the executive authorities of the Russian Federation subject will set a uniform rate and the operator's expenses on MSW treatment and recycling will not be considered. The regional operator will be required to conclude a contract for providing services with either managing board of the apartment buildings or with the premises owners directly.

Thus, all the contractual relations of the managing boards with the transporting organizations, landfills and other companies processing the living quarters waste must be closed before December 31, 2015, as from January 1st, 2016 the only allowable party in the relations associated with the MSW for the managing boards will be the regional operator – the unified organization, that will provide for the service for the MSW treatment and getting payment for that from the residents..

It is planned that the environmental tax funds will be used for the development of the industry.

The concept of **"environmental tax"** was introduced by the Federal law No. 458 starting from January 1st, 2015. The said payment relates to the non-tax revenues of the federal budget and must be paid by the manufacturers and the importers of the recyclable goods. The essence of the introduce changes is that manufacturers, the importers of the goods are required to provide for the recycling of the waste generated by their goods consumption and after these goods lose their consumer properties either by their own or by paying the environmental tax. The list of such goods, including the packaging, corresponding recycling standards, environmental tax rate for each group of the goods to be recycled after the loss of their consumer properties, and the procedure of the tax collection shall be established by the Russian Federation Government.

The funds received by the federal budget as the environmental tax will be distributed via the realization of the state programs of the Russian Federation in the form of grants to the Russian Federation subjects in the following areas:

- co-financing of the regional programs in the field of waste treatment and regional waste treatment scheme;
- covering of the expenses on the collection, transportation, processing, and recycling of the waste generated by the use of the goods;
- covering of the deficit of the funds received as a payment from population for the public services for the MSW treatment;
- the execution of the engineering survey and preparation of the project documentation for the construction of the facilities used for the waste recycling, waste neutralizing facilities, and the equipping of such facilities.

The said grants from the federal budget will be given in the manner established by the Russian Federation Government in proportion to the population living at the territory of the Russian federation subject.

In June of 2015 a meeting was held with the participation of the Russian Federation Government to establish the limits of the non-tax payments for the businessmen and organizations, where it was decided to introduce a **moratorium in respect of the non-tax payment** of the businessmen and organizations till January 1st, 2019. According to the results of the meeting a report No. DM-P13-48 gr of 01.06.15 was prepared with the instructions of the Russian Federation Prime Minister D.A. Medvedev, including the task to set the standard of the utilization for all categories of goods at the rate of 0%, excluding batteries, tires, paper, light bulbs as well as the groups of goods, which are already under utilization.

Since January 1st, 2016 the payers of the fee for the negative environmental impact (NEI) when placing MSW must have been the operators on MSW treatment, regional operators carrying out the placing of the MSW. Besides, the expenses on the NEI payment when placing the MSW must be considered while setting the rate for the MSW treatment operator, regional operator in the manner prescribed by the fundamentals of pricing in the field of MSW treatment. As we can see, the changes provided by the Federal Law No. 458 significantly limit the rights of the local autonomous bodies in the field of the waste treatment.

In the connection with the unfinished legislative base, in particular for the transition to the new system of MSW management, in December of 2015 the RF President signed the Federal Law No. 404 of 29.12.2015 "On Amending the Federal Law "On the Environment protection" and the Russian Federation Certain Legislative Acts". The Law No. 404 actually introduces a transitional period in respect of the municipal solid waste treatment till January 1st, 2017 (earlier the Law No. 458 provided for the introduction of the rules for the MSW treatment since January 1st, 2016). During the transitional period it is supposed to perform the selection of the regional MSW treatment operators and to introduce a new municipal service "MSW treatment". Until then the waste removal will be still carried out as the part of the housing service in the form of "MSW collection and removal".

However, the Law No. 404 did not cancel the effectiveness since January 1st, 2016 of the most part of the provisions of the Law No. 458. The terms "Municipal solid waste" and "MSW treatment" are now enshrined in law, a number of rules came into force, regulating the activity in the field of MSW treatment: it is provided for the creation of the regional operators at the territory of the RF subjects who will collect, transport, process, utilize, neutralize, and dispose the MSW; the peculiarities of the contract signing for the service MSW treatment, the types of activity and rates, that are to be regulated; the procedure of the rate state regulation, the peculiarities of the organization of the state control (supervision) in the field of the rates regulation; the term "waste accumulation" is corrected, which is now understood as a temporary wage storage (for a period of no longer than 11 months) with a view to its future use, neutralizing, placing, and transportation (earlier this period was 6 months); the other provisions regulating the MSW treatment activity are also provided for.

Thus, the most complicated tasks currently facing the Russian Federation subjects executive authorities are:

- to develop the territorial scheme of the waste treatment;
- to appoint the regional operators;
- to set the rate limit (the rates for the MSW processing, neutralization and disposal, and the uniform rate for the regional operator service of the MSW treatment);
- to approve the regional programs in the field of waste treatment;
- to sign the contracts;
- to fill up the deficit of the waste processing facilities.

The new waste treatment system (Figure 3) is similar to the modern regulation models in the field of heat supply, water supply and water drainage [3].

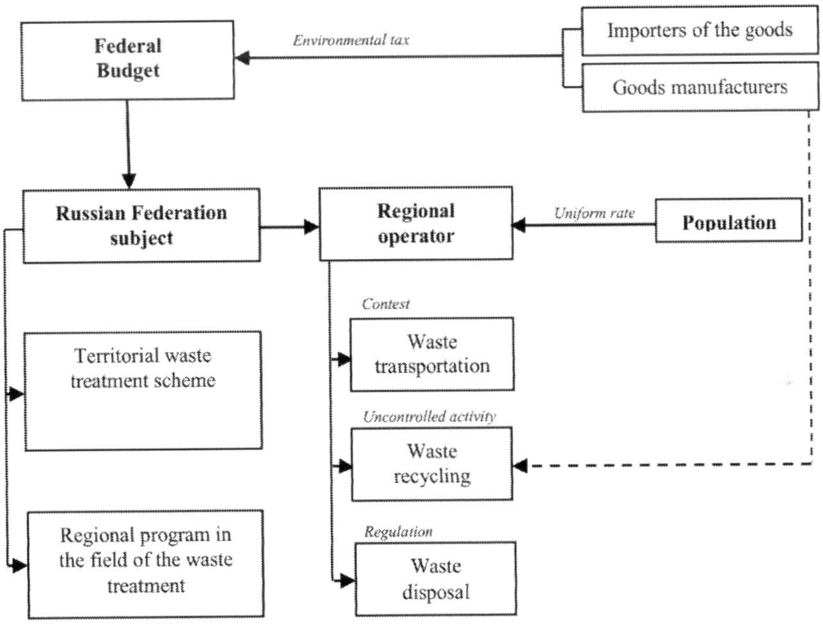

Fig. 3. New system of the waste treatment.

The acts of law of the Russian Federation subjects must regulate the relation connected with:

- the manner of development, coordination, and approval of the regional programs in the field of waste treatment;
- the content and procedure of the making agreement between the Russian Federation subjects executive authorities and the regional operators, as well as the conditions of the biddings for the carrying out the collection and transportation of the MSW;
- the procedure of the transfer to the local governments of the settlement or urban district of authority to establish the standards of the MSW accumulation (if necessary);
- the activity of the regional operators.

DOI: 10.1051/e3sconf/20160601006

3 The main provisions of the integrated waste management system in the Sverdlovsk region

Today at the territory of the Sverdlovsk region a difficult situation in the field of MSW treatment has arisen. The reforming of the MSW treatment activity and the selection of the optimal method of the waste utilization are the complex and multiple-factor task. The solving of this problem is based on the comprehensive approach. The comprehensive approach to the MSW management must consider economical, social, environmental, technical, and organizational aspects. All these aspects should be considered in combination, taking into account the specific conditions of each oblast (region). The waste integrated management (WIM) assumes that the approaches to the solving problem of MSW are selected depending on the specific local conditions and resources. While defining the aims of the MSW utilization program and planning the strategy it is advisable to have an idea about a certain WIM hierarchy. This hierarchy implies that first of all the measures for primary reduction of the waste should be considered, followed by the secondarily reduction: the reuse and recycling of the rest of the waste, and the last thing is the measures for utilization or disposal of the waste, generation of which could not be avoided and which are not recyclable.

In this regard, the following main principles of the waste integrated management can be emphasized:

1) MSW consists of various components, which should get the an individual approach;

2) all technologies and measures for MSW treatment should be developed in complex using their combination;

3) the territorial MSW treatment system should be developed considering the specificity of the local conditions and resources;

4) it is necessary to be based on a strategic long-term planning, which provides for the possibility to quickly and accurately consider the changes in MSW treatment, and to timely monitor and evaluate the results;

5) the participation of the municipals and all groups of population is necessary for the more effective solution of the MSW problems.

Considering the innovations in the MSW treatment system, the comprehensive approach to the waste management may be presented as following (Figure 4).

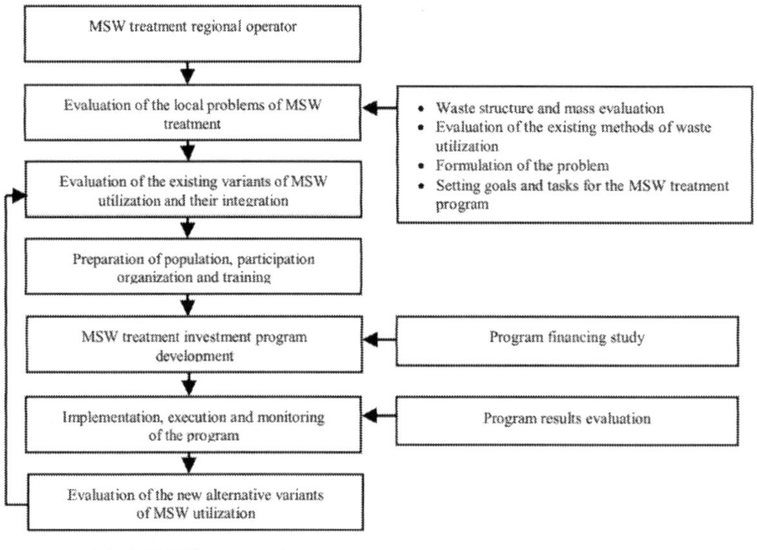

Fig. 4. The stages of the MSW integrated management.

In the Sverdlovsk region they have already started developing the main mechanisms of the organization system of the municipal solid waste processing [5]. Back in September of 2015 the agreement between the Sverdlovsk region government and the state corporation was signed to establish a joint venture for the municipal solid waste treatment (25% of its shares will belong to the Sverdlovsk region, the rest – to the company). The new structure will select the contractors working with waste both in general and on each technological stage. And this structure will accumulate the payments from population and organizations. The projects of modernization of the existing municipal MSW landfills and the construction of the new ones at the expense of the investors and coordination of their fulfillment by the attracted investors will be in the competence of the operator. The contests to select the investors are planned for 2016. The new organization will coordinate the work of all participants of this process: the managing boards collecting MSW, transporters, landfills owners and the others. The regional government expects that during several years at least 6 inter-municipal complexes will appear in the Middle Urals. The construction must start at the territory of the central industrial cluster of Sverdlovk region [5].

Amended and expanded by the authors SWOT-ANALYSIS of the scope of the municipal solid waste treatment at the territory of the Sverdlovsk region is given in Table 3 [6].

Table 3. SWOT-ANALUSIS of the scope of the municipal solid waste treatment at the territory of the Sverdlovsk region.

Strengths (S)	Weaknesses (W)
1. The business community's interest in the MSW treatment sphere development. 2. The presence of the companies recycling SMR. 3. The interest of the regional authorities, executive authorities of the Sverdlovk region, local autonomous bodies and population in MSW problem solution. 4. The presence of the municipal and state programs for the MSW treatment and preservation of the environment. 5. The executive authorities and local autonomous bodies conduct measures aimed at the gradual transition to the new form of management, which will provide for the stable functioning of the system in the field of MSW treatment. 6. The presence of the ground areas for the construction of the modern MSW landfills. 7. The presence of the MSW processing innovative technologies. 8. The presence of the Sverdlovsk regional cadastre of the industrial and consumption waste.	1. The low level of the environmental culture of the population. 2. The lack of logistical support of the MSW collection and utilization process. 3. The lack of the MSW landfills meeting the requirements of the environmental and sanitarian legislation. 4. The absence of the civilized MSW collection and utilization market. 5. Failure to comply with the environmental and sanitarian legislation while treating MSW. 6. The absence of the centralized MSW management authority in the region. 7. Ineffective accounting system for MSW generation, use, and utilization. 8. Poor organization of the MSW collecting in the rural communities and the private sector. 9. The absence of the financial ability to continue serving the sites by the managing companies, as the money from the residents for the "MSW treatment" service will be received by the regional operator. 10. The introduction of the transition period due to the insufficient workup of the legislative innovations. 11. The payment of the environmental tax is waived until January 1st, 2019.
Opportunities (O)	**Threats (T)**
1. The development of the MSW and SMR processing market. 2. Attracting the population to MSW problem solving. 3. The MSW separate collection introduction and	1. The deterioration of the ecological and sanitary-epidemiological status of the territories: the air, soil, and ground waters pollution. 2. The existing problem of the MSW treatment

organization. 4. The introduction of the innovative technologies for the MSW processing (more environmentally friendly). 5. Improving the legal framework in the field of MSW treatment. 6. Environmental upbringing and education, starting from the kindergarten children, schoolchildren, students of the educational establishments to the specialists of the enterprises and organizations, as well as state and municipal authorities. 7. The development of the public-private partnership, the maximum involvement of the private investors in the MSW treatment system. 8. The improving of the rate regulation system for treating MSW 9. The creation of the unified MSW treatment system. 10. The MSW problem solving using the program-target method. 11. The future availability of the information regarding the regulated activities in the field of MSW treatment.	leads to the social tension increase. 3. The overfill of the MSW landfills. 4. The emergencies in case of the MSW utilization technologies violation. 5. The negative effect of the selected technologies introduction. 6. The double payment for the said municipal service by the most part of the working population. 7. The impossibility to transfer some specific types of the MSW.

SWOT-ANALYSIS allowed the emphasizing of the main problems in the field of MSW treatment at the territory of the Sverdlovsk region. The necessity to solve the said problems requires the perfection of the mechanisms and methods of the MSW treatment at the territory of the Sverdlovsk region.

The most important tool for the solving the posed tasks will be the comprehensive approach to the MSW treatment. To evaluate the effectiveness of this approach one may use the target indexes. The main target indexes characterizing the MSW treatment activity are presented in Table 4 [5].

Table 4. Target indexes of the waste integrated management in the Sverdlovsk region.

Target index title	The target indexes values, progressive total				Basis value of the target index (at the beginning of the strategy realization) 2013
	Stage I (2013–2016)	Stage II		Stage III (2025–2030)	
		2016–2020	2020–2025		
The ratio of MSW processing to the total volume of its generation per year, %	15	35	55	80	12
The share of the buried MSW in the total generated volume, %	80	65	45	20	88
The share of MSW, passing trough the separate collection and assorting, %	20	50	70	80	18
The share of the secondary material sources, extracted from the MSW total volume, %	20	50	60	–	12
The share of the landfills meeting the requirements	14	50	80	100	13

of the environmental legislation, %					
The share of the abandoned illegal dumps, %	70	100	–	–	68
The share of the reclaimed objects for the MSW disposal facilities, %	20	50	75	100	10
The share of the stationary and mobile reception points for the secondary resources placed in operation, %	35	60	100	–	34
The share of the municipal formations, at the territory of which the sanitarian cleaning schemes were developed, %	100	–	–	–	65
The share of the MSW, used for the energy production, %	5	5	10	30	0

In conclusion it may be said that the realization of the waste integrated management in region is associated with the risk, which may impede the achievement of the set targets and results planned. Such risks may include: regulatory risks, macroeconomic risks, financial risks, organizational risks, social risks, natural risks, technical and innovation risks.

The measures of state regulation and management of the above risks able to minimize the consequences of the negative events and processes should include:
1) the creation of the effective control organization system for the Strategy fulfillment, as well as providing the stable development of the MSW treatment sphere at the existence of the said risks;
2) carrying out the planned checks of the nature resources using enterprises subject to the regional state environmental supervision;
3) annual monitoring of the environmental situation in the region;
4) the adoption of the normative legal acts, regulating the relations in the field of environmental protection and providing for ecological safety in the Sverdlovsk region.

Thus, WIM will allow the creation and providing for environmentally safe and economically effective system in the field of MSW treatment at the region territory.

References

1. State report "On the State and Environmental Protection of the Russian Federation in 2014", http://www.mnr.gov.ru/regulatory/list.php?part=1756
2. State report "On the Condition and Environment Protection in the Sverdlovsk Region in 2014", http://www.mprso.ru/gosudarstvennye-doklady-o-sostoyanii-i-ob-ohrane-okruzhaiushei-sredy-sverdlovskoi-oblasti
3. O. Shaposhnikova, Ecologist's Reference-Book, **8** (**4**), 48–54 (2015)
4. M. Lamihova, Ecologist's Reference-Book, **5** (**3**), 12–19 (2015)
5. The Sverdlovsk Region Government official web-site. http://www.midural.ru/news/list/document70278/
6. The Municipal Solid Waste Treatment Integrated Strategy at the Territory of the Sverdlovsk region till 2030. http://energy.midural.ru/images/Upload/822/doctbo/1259-%D0%9F%D0%9F.pdf

E3S Web of Conferences **6**, 01007 (2016) DOI: 10.1051/e3sconf/20160601007
ICSC 2016

Potential of Sverdlovsk Oblast and Yekaterinburg Environmental Sector

Irina Belik[1], and *Natalia* Starodubets[1a]

[1]Ural Federal University, 620002, Yekaterinburg, Russia

Abstract. The article discusses the problems associated with introduction of new administrative approaches to solid waste management, as well as the choice of priority directions in the development of the regional waste treatment schemes, which may increase the efficiency of the environmental sector in the city (region) and improve the ecology in general. The authors analyze the factors, which may lead to environmental deterioration and endanger the ecological safety of Yekaterinburg and Sverdlovsk Oblast. They also highlight the priority challenges, one of which is utilization of the biogas produced in the municipal solid waste landfills. In the context of the existing administrative districts, the paper analyzes the results of calculations of biogas production and emissions from the waste incineration plants in Yekaterinburg and Sverdlovsk Oblast. The authors offer an algorithm, which helps to evaluate the reduction of methane emissions to the atmosphere after introduction of a biogas combustion technology. The article also discusses the author's recommendations on how to justify the directions in the development of the regional waste management schemes, which will increase the efficiency of the companies operating in the field of waste management.

In compliance with Decree No. 1662-r of the RF Government dated 17.11.2008 one of the main targets for the long-term socio-economic development of the Russian Federation for the period up to 2020 is safety of the country's citizens and society. The transition of the Russian economy from the raw material export pattern to the innovative social-oriented type of development has been chosen as a leading comprehensive and coordinated strategy for achieving this goal. One of indispensable prerequisites for this process is the creation of an innovative and efficient environmental sector, which will meet the needs of the country's growing economy. The adoption of the Russian Environmental Protection Strategy for the period up to 2030 and Federal Law No. 458-FZ dated 12.29.2014 have become the most important steps along this way. In compliance with the last-mentioned law governing waste management sphere, regional schemes should be developed and regional operators should be appointed. It is obvious that realization of this goal may be achieved only after creation of a unified system of control over waste collection, sorting, storage and recycling (Fig. 1).

Regional operators must be appointed on the territory of Sverdlovsk Oblast as well as in other federal subjects of Russia. Their duties will include the development of principle

[a] Corresponding author: n.v.starodubets@gmail.com

directions in the field of waste management in the region and major cities. The following areas of focus have been determined for Sverdlovsk Oblast:
1) delivery of wastes to the recycling plants and landfill sites;
2) use of wastes as secondary raw materials;
3) use of biogas as a source of thermal and electric energy.

Figure 1 shows the municipal solid waste management system (MSW), which is the same for all the subjects of the Russian Federation. However, the environmental and economic peculiarities of each region impose their own specifics on the regional schemes and programs (local context).

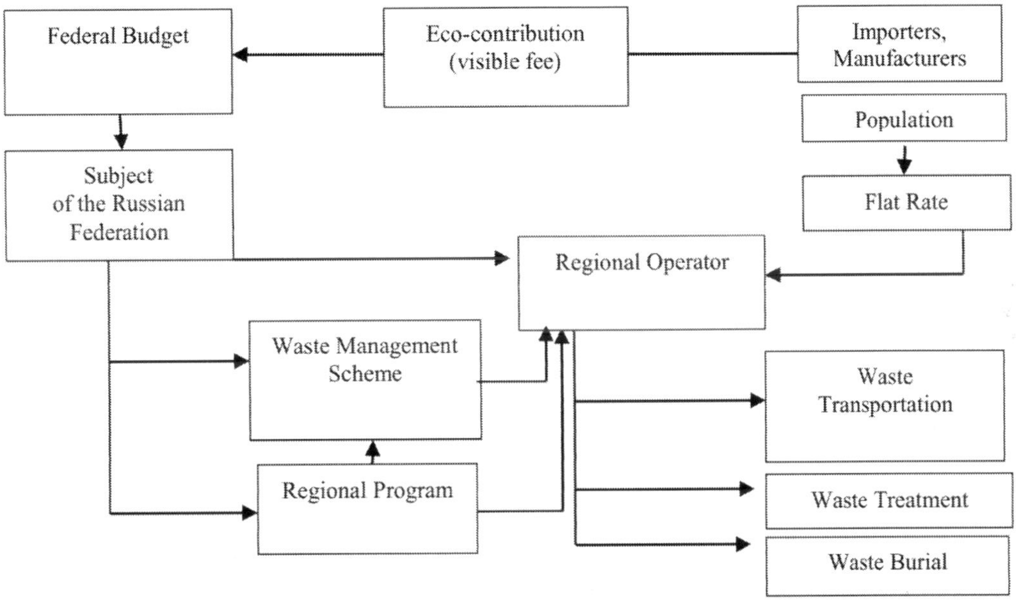

Fig. 1. Waste Management System

In addition, the developers of waste management schemes and regional programs need to create an adequate mechanism for assessment of the environmental effectiveness and economic efficiency of different business options and proposals related to recycling [1]. Consequently, it will be interesting to formulate an algorithm for calculating the volume of emissions from the biogas incineration plants.

Figuratively speaking, the process may be subdivided into five stages. The first stage is to calculate the volume of solid wastes disposed of at landfills. In Yekaterinburg and Sverdlovsk Oblast these calculations were made for each administrative district using the data presented by the Federal Service of State Statistics [2]. On average, per capita waste generation in the region is approximately equal to 400 kg/year. The population of Sverdlovsk Oblast is 4 320.7 thousand people. Table 1 shows the data on the population in the administrative districts.

Table 1. Population of Sverdlovsk Oblast [2]

Index	Total	Administrative District of Sverdlovsk Oblast					Yekaterinburg Municipal District
		Eastern	Southern	Northern	Gornoza-vodskoy	Western	
Number	4 320,7	490,7	620, 3	485,4	691,5	587,1	1 445,7

of residents as of 01.01.2014, thousand people (%)	(100%)	(11,4%)	(14,3%)	(11,2%)	(16,0%)	(13,6%)	(33,5%)
Number of municipalities	94	32	14	15	12	20	1

The annual volume of wastes generated in Yekaterinburg and in each administrative district is calculated on the basis of the data from Table 1 (population size) and Table 2 (waste generation).

Table 2. Waste Generation

Index	Total	Administrative District of Sverdlovsk Oblast					Yekaterinburg Municipal District
		Eastern	Southern	Northern	Gornozavodskoy	Western	
Municipal solid waste generation, thousand kg/year	1728280	196280	248120	194160	276600	234840	578280

The second stage is aimed at determining the amount of biogas generated at landfills. For this purpose, it is necessary to analyze the waste composition since it may vary greatly depending on the country, region or population cluster (town, village). The composition of solid wastes is affected by a number of factors, the most important of which are the consumption patterns, the average household income, the type of dwelling (a private house or an apartment building), etc. Thus, EuroStat reports that in the developed countries organic/food/green wastes account for about 55% of the total waste volume. Other fractions such as paper and paperboard constitute 23%, while plastic and metals are 8% and 4% respectively.

The annual volume of biogas generated at landfills under the study was evaluated using "the method of calculating the measurable parameters of pollutant emissions into the atmosphere from the landfills of solid domestic and industrial wastes." Specific biogas yield during the period of its active stable generation in the process of methane fermentation is calculated from equation [3]:

$$Q = 10^{-6}\,(100 - W)\,R\,(0{,}92\,F + 0{,}62C + 0{,}34\,P), \quad (1)$$

where Q is the specific biogas yield during the period of its active generation, kg / kg of wastes;
R is the content of organic components in wastes (assumed to be – 55%);
F is the content of fat-like substances in organic wastes, (assumed to be – 2%);
C is the content of carbohydrate-like substances in organic wastes (assumed to be – 83%);
P is the content of protein substances in organic wastes (assumed to be – 15%);
W is the waste humidity (assumed to be – 47%).

Taken into account the aforementioned assumed values, the total biogas content in 1 kg of wastes is equal to 0.170236 mg / kg.

An annual quantitative yield of biogas per one ton of wastes is calculated by formula [3]:

$$P_y = (Q{*}10^3)\,/\,t_{co\text{-}dig}\ kg\,/\,t \text{ wastes a year} \quad (2)$$

where $t_{co\text{-}dig}$ is the period of full scale co-digestion of organic wastes over the period of one year. It is calculated from empirical formula [3]:

$$t_{co\text{-}dig} = 10248 / (T_{warm} * (t_{av.\ warm.})^{0,301966} \quad (3)$$

where: $t_{av.\ warm}$ is the mean value of the monthly average air temperatures ($t_{av.\ month}$) around the landfill of municipal and industrial solid wastes (MSW and ISW) during the warm period of the year, $t_{av.\ month} > 0$, (assumed to be + 11°C);
T_{warm} is the duration of the annual warm period in the area around the landfill of the municipal and industrial solid wastes, in days (assumed to be 200 days);
10248 and 0.301966 are the specific coefficients, which describe the biothermal decomposition of organic matter.

Taking into account the assumed values, the biogas yield for the year amounted to 6.853423 kg/t of waste. On the basis of the data on the annual biogas generation and the population of Sverdlovsk Oblast the authors calculated the volumes of biogas in the landfills of every administrative district (Figure 2).

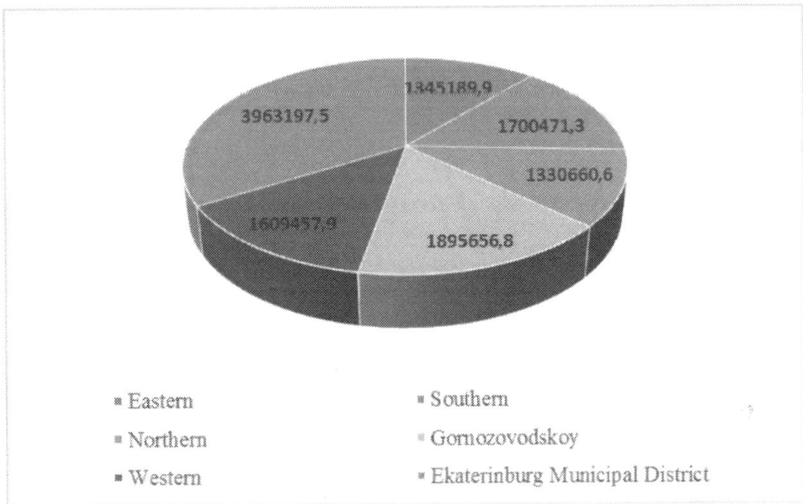

Figure 2. Maximum biogas yield in every administrative district and in Yekaterinburg, kg/year.

The aim of the third stage is to determine the emissions connected with biogas generation. Biogas is a mixture of methane, carbon dioxide, nitrogen dioxide and oxygen. Depending on the landfill age, biogas composition may vary, that is why the data from Table 3 should be used to calculate the content of each gas [4].

Table 3. Biogas composition under different landfill conditions [4]

Biogas type*	Methane,%	Carbon dioxide,%	Oxygen, %	Nitrogen, %
1	55	45	–	–
2	40	30	6	24
3	45	35	1	18
4	35	30	5	30

*Note): Type 1 – pure biogas produced under anaerobic conditions, type 2 – the oxygen and nitrogen content in the landfill gas is approximately the same as in the atmospheric air; type 3 – the oxygen from the air over the landfill is used in the microbiological process; type 4 – a combination of 2 and 3.

The volume of gaseous pollutants in the landfill gas, emitted into the atmospheric air, is defined using the data from Table 4 (the combined type of gas generation). The calculated volumes of gaseous pollutant emissions into the atmosphere of each administrative district in Sverdlovsk Oblast and Yekaterinburg are shown in Table 4.

Table 4. Volumes of gaseous pollutant emissions connected with different biogas components

Name	Total, t/year	Administrative Districts of Sverdlovsk Oblast					Yekaterinburg Municipal District
		Eastern	Southern	Northern	Gornozavodskoy	Western	
Methane (CH4)	4145622	470816	595165	465731	663480	563310	1387119
Carbon dioxide (CO2)	3553390	403557	510141	399198	568697	482837	1188959
Oxygen (O2)	592231,7	67259	85024	66533	94783	80473	198160
Nitrogen (N2)	3553390	403557	510141	399198	568697	482837	1188959
Total	11844634						

However, biogas combustion prevents methane emissions into the atmosphere. Methane has the greatest impact on the greenhouse effect, which is 21 times stronger than that of carbon dioxide. Moreover, methane remains in the atmosphere for an average of 12 years. As we know, methane capture is the best short-term way to prevent global warming. Climate change is manifested in the increased number and strength of different weather events: floods, droughts, heat waves, sharp frosts, strong winds, heavy snowfalls, etc.

The purpose of the fourth stage is to investigate the energy potential of wastes and to offer the options for their use as fuel. The biogas, produced in the landfills of municipal solid wastes (MSW), contains a significant amount of methane, which may be used for energy generation. It should be borne in mind that though, on the one hand, biogas combustion significantly reduces the volume methane emissions, on the other hand, it increases the discharge of harmful exhaust fumes from burning in a CHP plant (Table 5).

Table 5. Greenhouse gas emissions from biogas combustion in CHP [5]

Biogas consumption, thousand m^3/year	Electric power production, thousand kWh /year	Greenhouse gas emissions, t/year		
		CO_2	N_2O	CH_4
345	828	414,2	0,001	0,038
690	1656	828,3	0,002	0,076
1380	3311	1656,7	0,003	0,152
2129	4967	2556,0	0,005	0,234
2759	6623	3313,3	0,067	0,364

However, the use of the latest technology available (e.g. Japanese Patent H 05-245339 or a distillation column designed by Russian engineers) offers ways for reduction of carbon dioxide emissions into the atmosphere associated with combustion of biogas.

Stage 5 is used to calculate the amount of individual components (methane, carbon dioxide) in the greenhouse gas emissions from combustion of biogas. The volume of greenhouse gases emitted during burning of 1,000 tons of wastes is determined with the use of the interpolation method on the basis of the data presented in Table 5. The volume is calculated by the following formula:

$$18 + ((6623 - 4500) * (34 - 18)) / (9900 - 7500) = 24290{,}37 \text{ t/year}. \quad (4)$$

According to the obtained results, the emissions of individual components of the greenhouse gases during combustion of 24 290.37 tons of biogas will be as follows: 3 313.3 tons of carbon dioxide and 0.364 tons of methane per year. Calculations of biogas volumes generated at landfills indicate that despite the use of combustion technology, there is a significant reduction in emissions of methane into the atmosphere of Sverdlovsk Oblast and Yekaterinburg. Table 6 shows the potential for reduction of harmful emissions after construction of incineration plants, which will burn 24 290.37 tons of biogas a year. Such plants are planned to be located in every administrative district of Sverdlovsk Oblast and in Yekaterinburg.

Table 6. Potential for reduction of harmful substances

Index	Administrative District of Sverdlovsk Oblast					Yekaterinburg Municipal District
	Eastern	Southern	Northern	Gornozavodskoy	Western	
Methane (CH4)	412551	536900	407466	605215	505045	1328854
Decreased by, %	12	10	13	9	10	4
Total	9,7 %					

The calculations have shown that the use of biogas generated at landfills is technically possible (since this gas has a high energy potential) and environmentally feasible, because it allows to significantly reduce methane emissions. The results obtained following the reduction of greenhouse gas emissions from combustion of biogas may be applied to the regional waste management schemes, introduced starting from 2015 in accordance with Federal Law No. 458-FZ, as a priority area of the environmental sector development. This is especially important for urban territories, since there is a progressive growth of solid wastes accompanied by the negative tendency in waste generation rate, which exceeds the population growth rate. The technical possibility for burning of biogas to generate heat and electricity leads to a decrease in the volume of emissions into the air and allows to speak about the improved economic efficiency of waste management. The necessary financial resources for the implementation of this technology can be received from the environmental fee (see Fig. 1). However, to make this scheme work, the local government together with the regional operator need to additionally create a secure system of waste collection and transportation on the municipal territories (the city), as well as the conditions for safe waste disposal in the landfills, which should be built in accordance with the regulatory requirements.

References

1. I. Belik, N. Starodubets, *Conference Proceedings 8th Annual Conference of the EuroMed Academy of business*, 2468–2472 (2015)
2. Official website of the Federal State Statistics Service of Sverdlovsk Oblast, http://sverdl.gks.ru/

3. Method for calculation of measurable parameters of pollutant emissions into the atmosphere from the landfills of solid domestic and industrial wastes, http://www.waste.ru/modules/library/singlefile.php?cid=1&lid=6
4. Guidelines for calculation of greenhouse gas emissions from solid waste landfills, http://climatechange.kz/docdowlon/metodik/othod/W1.3.pdf
5. O.V. Masleeva, G.V. Pachurin, Fundamental research, **6-1**, 139–144 (2012)

E3S Web of Conferences **6**, 01008 (2016) DOI: 10.1051/e3sconf/20160601008
ICSC 2016

Forming the social optimum of passenger flows in urban agglomerations in Russia

Igor Mayburov[1a], and *Yulia* Leontyeva[2]

[1]Graduated School of Economics and Management, Ural Federal University, 620002, Yekaterinburg, Russia; Far Eastern Federal University, 690950,Vladivostok, Russia
[2]Graduated School of Economics and Management, Ural Federal University, 620002, Yekaterinburg, Russia

Abstract. The article asserts the role of transport policy in making a city comfortable for living and tourism. A theoretical framework is built for reaching the social optimum of passenger flows. The authors argue that reaching the social optimum is only possible by implementing a system of measures that would encourage the use of public transport and, in parallel, by implementing measures that would discourage people from using private cars. Substantiation is provided for an idea that an unregulated urban transport system is less effective than a regulated transport system that provides targeted incentives and deterrents for various personal preferences of city dwellers in order to reach the social optimum. The authors that variable taxes have the strongest effect on travel behavior. Comparative analysis is done of the specific features of various types of variable transport taxes. The authors propose measures for encouraging the use public transport and calculate the fiscal impact of introducing a distance-based toll levy in Yekaterinburg. The authors argue that a charge that vary by the time of day will make it possible to ensure a more even use of roads and parking space.

Introduction

Creating rationally structured transport systems is critical for improving the quality of life in modern urban agglomerations. The image of urban agglomerations and the quality of life there depends on how this issue is addressed.

Cities that commit themselves to encouraging the use of private cars have to expand their road networks at an accelerated pace. In densely built-up residential areas, such expansion is inevitably associated with changes in the structure of urban land. Roads take over land designated for other purposes, with land for recreational use (parks and boulevards) being the first one to go. When the capacity of the road network grows, privately owned vehicles start to be used with an increasing intensity. As a result, using a car in the city becomes even more attractive an option; the number of rides goes up; congestion reappears, spurring the need for more and wider roads. This spiral kept unwinding in the majority of US cities in the second half of the 20th century.

[a] Corresponding author: i.a.mayburov@urfu.ru

Cities that adopt the policy of encouraging greater use of public transport have achieved completely different results. Many urban agglomerations in Europe integrated all modes of public transport into a single network and focused on the development of rail transit, thus achieving a shift in individuals' preferences to public transport. As a result, the cities were able to create an environment that is comfortable for city dwellers and tourists.

Transport policy in most Russian urban agglomerations is yet to take shape, so the problem of setting the right priorities is of prime importance to Russia.

1 Theoretical framework of the social optimum of passenger flows

The search for a social optimum in traffic allocation was pioneered by John Glen Wardrop [1]. In 1952, he studied an equilibrium distribution of public and private transport flows within a section of a road network. The point of equilibrium was determined by comparing total disutility (total costs) of all travelers. Wardrop proved that when each road user chooses their preferred means of transport, this choice is not socially optimal (Figure 1).

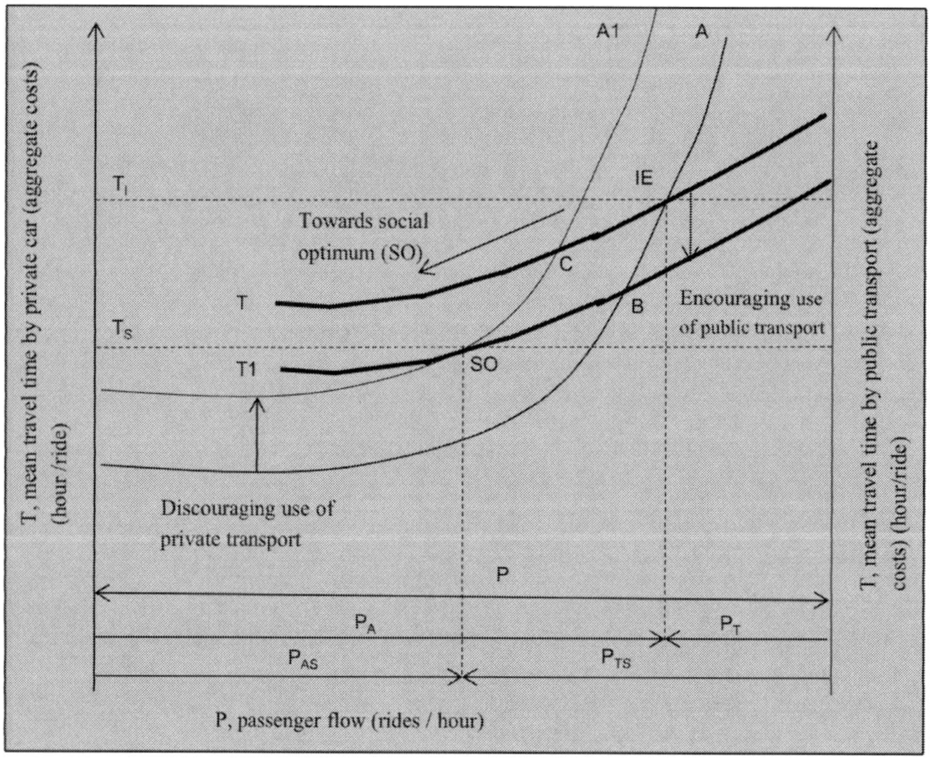

Fig. 1. A combination of transport policy measures aimed at shifting equilibrium of individuals' preferences towards the social optimum (SO)

The passenger flow, P, moves along the abscissa to the right for the A-curve (total expenses of a motorist) and from right to left for the T-curve (total expenses of a public transit passenger). In an ordinary situation, the dependences of aggregate costs on the density of the traffic flow for private and public transport intersect at point IE.

The point is the equilibrium point of individuals' preferences. A number of city dwellers (PA) decide to travel by automobile, while another part (PT) of them prefer public transport

(PA>PT). The aggregate expenses of all city dwellers are given by the area of the triangle restricted by line T1.

V. Vuchic argues that individuals' preferences as to the means of transport are pretty stable [2]. Despite any swings of the preferences, urbanites will eventually return to the point of equilibrium. In order to move the equilibrium point to the left towards the social optimum and ensure the stability of the new combination of individuals' preferences it is necessary to simultaneously take a set of measures. It is necessary to implement incentives encouraging the use of public transport. At the same time, it is necessary to adopt measures to discourage the use of private transport.

If any of the sets of measures is adopted in isolation, the structure of the passenger flow will change insignificantly. For example, if incentives are provided for the development of public transport only, the T-curve will be shift to the position T1, while equilibrium will move to the point B. The time of travel will change only a little and there will be a certain growth in the number of public transport users. If only measures discouraging the use of private cars are introduced, effects will be insignificant either, with the A-curve moving to the position A1. Equilibrium will move to the point C.

Only combined adoption of all measures of transport policy ensures a shift of the both curves, while equilibrium moves to the social optimum point (SO) that has a completely different set of parameters. There, passenger flows are almost equally divided between public and private transport (PA~PT), while aggregate expenses of all urban dwellers who use both private and public transport are significantly reduced. The social optimum can, therefore, be defined as the point SO where total expenses of all road users are minimized.

A city public transport system that governs itself and functions in accordance with the laws of the market ensures equilibrium of individuals' preferences at the level IE. Such a system is less effective in all cases and generates bigger negative effects than a regulated transport system that provides targeted incentives and discourages various preferences towards the social optimum SO.

Specific ways that each urban agglomeration works out in order to shift the curves of preferences for private and public transport towards the social optimum should become the foundation of a long-term transport policy.

2 Discouraging private car use

Such disincentives should be introduced in the form of a set of monetary measures that increase car owners' expenses on car travel and make the use of a privately owned car less attractive.

Travel behavior of a city dweller will be a derivative of the price of the planned trip. In addition to the price the city dweller will make his or her choice by taking into account such factors as time and convenience.

The price of the planned trip will be the key choice factor, so taxes and charges levied upon the trip will have the greatest influence on changing the travel behavior of car owners [3, 4].

Transport taxes and charges are not essentially homogeneous; they have different impacts on the behavior of car owners. Similarly to fixed and variable costs, it is useful to divide mandatory transport charges into fixed and variable ones.

2.1 Fixed transport taxes and levies

Tax rates are not pegged to the intensity of car use. They are paid on a one-time basis when a car is purchased and then on a regularly basis regardless of how often the car is used. In Russia, fixed transport taxes include VAT and a stamp duty on newly purchased cars, registration fees, the car recycling fee, and annually paid vehicle tax. The size of the taxes is a determining factor in the affordability of a car [5]. If fixed taxes are fiscally insignificant, (less than 10 to 15 percent of the cost of owning a car), larger groups of the population can afford to own a car. This increases car ownership rates among city dwellers and brings about serious negative transformations in the urban environment. Higher fixed taxes (25 to 30 percent of total costs of owning a car) is an instrument of keeping some city dwellers (usually those on a low income) from owning a car. At the same time, fixed taxes have a neutral effect on the intensity of car use. They influence long-term decisions as whether to own a car, but have not impact whatsoever on people's travel behavior.

2.2 Variable taxes and levies

These are taxes on car use. They are designed to compensate for negative externalities that are generated by the running of automobiles. High rates of variable taxes have the greatest effect on people's current travel behavior [6]. Urban agglomerations in Europe actively use variable taxes such as fuel excise duties, toll charges, vignettes, congestion charges, distance-based electronically collected tolls, toll lanes, paid parking (Table 1).

Table 1. Comparative analysis of variable taxes

Tax (levy)	Description	Use limitations	Supplementary effects
Fuel duty	Surcharge on fuel price	Impossible to take into account the local transport infrastructure	
Vignettes	Paid when a vignette is bought for a limited period of time	Unrelated to intensity of car use	Absence of additional enforcement costs
Congestion charge	Paid upon entering a certain area within a specific time frame	Imposes additional enforcement costs of setting up collection checkpoints	It is possible vary the charge depending on time of day or week (by excluding weekends or off-peak hours)
Road toll	A one-time payment that depends on the covered distance	Additional costs of setting up toll collection checkpoints	Can vary depending on time of day
Distance-based electronically collected toll	Distance-based payment debited from an in-vehicle device	Substantial one-time investment; vehicles need to be equipped with onboard payment collection devices	Toll can vary depending on time of day Can be used for collecting payment for and control of paid parking Can be used for speed limit enforcement
Toll lanes	Payment debited when the car crossed into a toll lane; does not depend on the distance covered	Additional cost of control equipment	Can vary depending on time of day

In Russia, only fuel excises have found wide application, while paid parking and road tolls are only starting to be introduced. The process, however, is extremely slow and is meeting strong opposition from car owners. There are not examples to be cited of congestion charges and distance-based road tolls in Russian urban agglomerations. The fiscal burden of fuel excises is insignificant. We have calculated that they do not exceed

4 percent of the costs of car ownership. Such a light fiscal burden does not change the transport behavior of city dwellers.

Russian urban agglomerations must develop new fiscal instruments that would meet the following fundamental requirements: (1) toll revenues must be spent strictly on the objectives of transport policy in the urban agglomeration where the road system was used; (2) tolls must be levied according to the distance travelled and be related to the type of the road (federal, regional or municipal); (3) the charge should vary depending on the day of the week and the time of the day (the highest during peak hours and a minimal charge during the rest of the day); (4) the minimal charge should match the marginal costs of road use, while the maximum rate should also include a surcharge. The surcharge would balance demand for the road network and its capacity. Such an approach to toll setting is outlined in the 2004 recommendations of the Council of the European Union and has already been successfully tried and tested.

Distance related electronic tolling meets the requirements best. It can be considered as a promising fiscal instrument for Russian urban agglomerations.

3 Incentives for public transport use

Public transport use should be encouraged with the help of a set of measures that alleviate the disutility of travelling by public transport. The disutility could be reduced by increasing the speed of transporting passengers to their destination, offering alternative modes of public transit, making public transport safer and more comfortable; introducing unified fares and universal travel cards; streamlining and cutting transfer time during intermodal journeys [7].

In Russian urban agglomerations the disutility of public transport can potentially be mitigated by: (1) identifying sources of long-term investment; (2) accelerated development of all means of rail transport with dedicated lanes; (3) consolidating the ownership of public transport and building a unified municipal public transit system with a single owner who is capable of aligning the timetables of all means of transport and implementing unified transport logistics; (4) adoption of long-term planning and involving people living in the agglomeration into the planning process..

4 Calculation of fiscal impact of distance-related charge in Yekaterinburg

For the calculations, we used financial indicators of a new electronic road pricing system that is being introduced in Singapore. The system offers additional advantages such as total speed and parking control in the city and the option of issuing parking and speeding tickets.

Table 2. Financial dimension of introducing an electronic toll system in Yekaterinburg

№	Indicator	Value	Note
1	Total cost of introducing an electronic toll system		Including the installation of a control system and equipping vehicles with onboard electronic units
	million dollars.	300	
	billion roubles	21 000	US dollar to Russian ruble exchange rate set at 1 to 70.
2	General characteristics of city traffic flows		
	Daily mileage of an automobile, km/day	38	According to a survey conducted by Klakson newspaper

	Number of automobiles, thousands.	451.7	In Yekaterinburg
	Dailycity transport mileage, thousand km/day,	17 165	
	Annual city transport mileage, million km/year	6 265.2	
3	Estimated payback period, years	5	It is advisable to estimate a minimum payback period because technological solutions quickly become obsolete
4	Annual toll revenue ensuring payback period, million roubles/year	4 200	
5	Minimal toll charge, roubles/km	0.67	
6	Daily toll to be paid by one motorist, roubles/day	25.46	
7	Annual toll charge per car owner, thousand roubles/year	9.3	
8	Daily petrol expenses (with average fuel consumption being 8 ltr / 100 km and petrol price of 35 roubles per litre), roubles/day	105	
9	Annual petrol expenses, thousands roubles/year	38.3	

The calculations show that the minimum daily cost of using a car in the city will be 25 roubles, which is less than 25 percent of what the car owner spends on petrol. Additional revenues are guaranteed from freight vehicles. It is possible to vary the charge depending on the time of day, thus encouraging a more even use of the road network and parking space.

Conclusion

Russian urban agglomerations must adopt radically new approaches to long-term transport planning. It is necessary to introduce programs of public transport development and rail transit development in the first place. It is also necessary to adopt fiscal instruments that discourage the use of private transport in urban agglomerations. It is necessary to develop modern systems of distance-related toll. There is good potential for an electronic toll collection system that has the option of contactless pay-as-you-go payments.

References

1. J. G. Wardrop, Proceedings of the Institution of Civil Engineers. **1**, 325–362 (1952)
2. V.R. Vuchic, *Urban public transportation: systems and technology.*(Englewood Cliffs, NJ, 1981)
3. L.N. Lykova, Journal of Tax Reform,**1**, 76–89 (2015)
4. S.H. Burrington, *ConservationLaw Foundation (CLF)*(Boston, 1994)
5. The European Electronic Toll Service (EETS). Guide For the Application of the Directive on the Interoperability of Electronic Road Toll Systems. (Luxembourg, 2011)
6. I. Mayburov, Y. Leontyeva,*8th Conference of the Euromed Academy of Business. Readings book proceedings*, 1529–1541 (2015)
7. A. Jonathan,*Congested and Nowhere to Go: Congestion, Road Infrastructure, and Road Pricing in MetroVancouver* (2013)

E3S Web of Conferences **6**, 01009 (2016) DOI: 10.1051/e3sconf/20160601009
ICSC 2016

Green economy as a vector of building and development of smart cities

Valery Anufriev[1a], *Elena* Anufrieva[1], *Lidia* Petrunko[2], and *Aytkali* Kaminov[1]

[1] Ural Federal University, Higher School of Economy and Management, 620002, Yekaterinburg, Russia
[2] Ural Centre for Energy Efficiency and Environment, 620049, Yekaterinburg, Russia

Abstract. The vector of smart city development is power-efficient and low-carbon policy (strategy) within the framework of the green economy. Sustainable development and green growth of cities can be provided by active growth of the human capital owing to creative, innovative ideas, approaches and solutions. Idea creators should be prepared in the best green gymnasiums, universities, academies. Energy saving and energy efficiency, besides creation of conditions required for innovations in the main parts of economy, will considerably reduce the risks of climate change and natural and man-made disasters related to it. It will introduce considerable additional contribution into the development of smart cities.

Introduction

More than 74% of the Russian population live in cities today. In this connection the sustainable development of Russian economy greatly depends on the level of city development and the quality of the human capital of urban population [1]. The modern city does not result only from its architecture and infrastructure. Thus the anthropogenic load on the planet due to the impact of the modern resource economy has brought about considerable overshoot in adaptation potentials of nature and contributes into the climate change. Nowadays the world economy consumes by one and a half more natural resources than the planet can reproduce. The situation is aggravated by the increase in the amount of domestic and industry wastes. The Russian economy is much more power-consuming than that of developed countries. And all of this is intensified by the influence of natural and manmade disasters related to emissions of greenhouse gases from burning hydrocarbon fuels, from landfill CH_4.

There is no alternative to transition to the green economy. More than 150 countries have changed or are changing over to the low-carbon economies. According to [2] in the beginning of 2015 "at least 164 countries had renewable energy targets, and an estimated 145 countries had renewable energy support policies in place".

One of the main trends of the green economy is its energy efficiency that has great difficulties in building into the economies of Russia, the Republic of Kazakhstan, Ukraine and some others post-Soviet states.

[a] Corresponding author: v.p.anufriev@urfu.ru

Smart city

A smart city needs smart decisions, that lead to a new level of development. A smart city also needs innovation techniques providing economical and ecological utilization of the city infrastructure, which creates the modern level of life quality. "Smart city means smart management, smart habitation, smart people, smart environment, smart economy, smart transport" [3]. The conception of smart cities is being implemented actively, for example in Songdo City, South Korea. Construction of new smart cities is planned in China and India, the USA and the United Kingdom, Spain and Brazil. In Moscow's Expo Center there is a smart city prototype [4].

In our opinion "smart" city is a city where power-efficient and low-carbon technologies are used at both industrial enterprises and urban systems of life activity. It is the city that is served by green energy (by sources of renewable energy – the sun, wind, water, thermal pumps etc.) and green transport (electric vehicles, public transport facilities (better rail-tracked ones)), with maximum quantity of green zones. In smart cities traffic jams are struggled not only at the expense of multilevel garages, but by commissioning safe walking tracks and bikeways, then the most part of citizens would use bikes making better their health as well as the environment of the city. It is easy to determine how much any city is polluted, it is enough to look at the verdure in it. The exhausts of cars and industrial enterprises creates coating on plants and they die slowly. Verdure is the best detector of the ecological situation in cities. It is motor cars and industrial enterprises in cities that impacts the environment considerably.

It is impossible to create a smart city by the efforts of ordinary citizens only. More often it is the initiative of government or local authorities. Is it possible to change to sustainable green city development without the purposeful national green policy and economy? The answer that is staring us in the face is no. The experience of smart cities in Sweden, Canada etc. shows that all the necessary services for citizens should be in the area of 10–15 minute walk [5]. According to the strategic development plan "Vision 2030" the main targets for further improvement of the city were established: power effective transport, use of environmentally friendly construction materials, use of clean energy, rational use of the ground and water, waste processing with minimum influence on environment, etc. In addition to it in Stockholm the program on greenhouse gases emission reduction and climate change is realized [6].

And all this can be achieved only by careful planning of urban environment and considered introduction of green technologies.

Human capital

In a smart city (country) smart and industrious people live, they are able to study under anybody and to accept any help, talented people are the most valuable property in a state where the government is honest and competent and it is able to plan, to pose large-scale economic tasks for a long period of time and correct them depending on the situation, play an important role in investment attraction and innovation implementation [7].

It is such kind of people that can be attributed not to human factor but to human capital which will be effective in both exploitation of disappearing natural recourses and utilization of swiftly increasing space, industrial, and domestic wastes.

Smart people, probably, should live in smart houses. What is a smart house? It is a power-efficient house, with least emissions of greenhouse gases and pollutants, their roofs are decorated with a green loan or flower garden. It turns out that soil on a roof is excellent heat insulator, and that loft is the most energy wasteful premise. Elements of public infrastructure are controlled remotely though computer or mobile phone, that is at a distance of thousand kilometers the houseowner can adjust the temperature and humidity of air in the apartment, control over its security and air it as required.

Authors consider green technologies as innovations not only in technology but in law, insurance, taxes and other fields. The innovations promote green (smart) economy, the matrix of all the innovations. To create the possibility for such transition a new kind of specialists is required. Specialists in green economy and smart citizens should be educated in smart universities, that is green universities. We have already had such universities in Russia, MSU, MGIMO and others are among them. In Ekaterinburg such title can be given to UrFU, as well as USMU, USFEU, USUAA, USAU.

For that continuous teaching green economy is required from a green gymnasium (kindergarten) to a green academy created on the basis of several green universities, with training courses on green economy given by teachers of the universities who specialize in a certain discipline, that considerably increases the course rating. It can be the second diploma within the framework of distant education on Sustainable Development and Green Growth offered by universities of BRICS countries. One of the main tasks of the universities is to give their graduates wide world view and to teach them thinking constructively, as well as to find talented specialists.

Behind every idea and innovative decision there is an individual (or a group), the idea creator. Naturally in the absence of such people the ideas providing the continuity of opportunity growth and the continuity of development are lacking. But history shows that the people were, are and will be. Exactly the people with their theoretic investigation and scientific and technical developments create prerequisites for choice of ways for further continuous development of society. Therefore the question is how to reproduce and form people who are able to generate and implement the ideas of continuous society development.

Creativity in economy is any process of search and implementation of solutions to complex tangled problems. The process of search and implementation of solutions to various economic, energy, climate, financial, social, legal, political and other problems is a creative one [8].

Green economy

Green economy (by authors' definition) is a system of economy management when economic growth is achieved owing to rational use of energy and natural resources, concomitant reduction of emissions of greenhouse gases and pollutants, as well as increase of the human capital significance (knowledge, creativity, culture).

The transition to the smart green economy takes place because the growing human capital, using their creative ideas and such scientific directions as neuroeconomics[b], will help to induce innovative power efficient low-carbon, and nature protective green technologies. The green technologies will provide the economic green growth in the main areas of economy and will give the following additional results:

- prevented environmental damage owing to reduction of pollutants,
- reduced damage from climatic changes and reduced risk of natural and manmade disasters owing to decreasing emissions of greenhouse gases.

We overlooked the moment when such terms as smart networks, smart houses and cities came into wide use, but our great compatriot academician V.I. Vernadsky anticipated

[b] Neuroeconomics is an interdisciplinary scientific direction on the border between psychology, neurobiology and economics theory. It deals with the study of the decision making process while choosing alternatives, distributing risks and remuneration. To study human brain the science uses economic models, and to create the economic models it uses the achievements of neurobiology.

the noosphere (the sphere of intellect) coming, when the thoughts, plans and business become reasonable. And we can suppose that power efficient low-carbon green economy is a predecessor of the noosphere economy [9, 10].

Green economy is the economy of the future because of its innovative character (40% of all innovations are implemented within its framework). Green economy is an antirecessionary economy, for example, South Korea accepted the idea of the economy as the national idea during the crisis of 2009 and become one of the first countries emerging from it [11]. And because of the abovementioned factors green economy has no alternatives.

The situation with fuel and energy balance of the countries changes very quickly at the expense of renewables and other types of green power. So, according to the International Energy Agency [12] the share of use of renewables in the world will increase from 22% in 2013 to 26% in 2020.

These years development of such branches as coal – and petrochemistry is planned. At the same time apart from use of coal as a fuel such innovative technologies as combustion of low-grade coals in the pseudo-liquefied layer, receiving of liquid fuel and synthetic gas from coal will find application.

Efficient energy consumption results not only in the consumption reduction but in the reduction of pollutants and emissions of greenhouse gases and therefore to reduction of the influence on climate changes: natural and manmade disasters, the damage of which is increasing annually.

Natural and manmade disasters and cataclysms increasing in frequency recently are related to the global climate changes in many respects. Retreat of the permafrost accompanied by worming in the northern regions of the country results in destruction of oil and gas pipelines, piled house footing, and manmade disasters. For example CEO UNILIVER official stated that the company loses about $330 million per year because of natural cataclysms due to climate changes [13].

Climate changes, and global warming in particular, depend on the increase of GHG emissions into the Earth's atmosphere (the main of them is CO_2). The gases are the result of hydrocarbon fossil fuel (coal, oil) burning. One of the key issues of the modern world economy is reduction of consumption of oil and coal as a fuel and transition to power-efficient and low carbon green power engineering an economy.

Greenhouse emission in Sverdlovskaya Oblast

In Russia the transition to power-efficient low-carbon development began in the first half of 1990s and moves forward with great difficulties. Universally recognized indicator of the power-efficiency is the amount of greenhouse emissions. Figure 1 shows the results of inspection of greenhouse gases in Sverdlovskaya Oblast, the forecast was made up to 2020.

The data obtained show that the emission of greenhouse gases in Sverdlovskaya Oblast from 2017 to 2018 does not fit the parameters marked by the President of the Russian Federation. It means that the energy-consuming enterprises of the region should quickly develop a program on reduction of greenhouse gases, coordinated with the regional energy saving program taking into account the level of greenhouse gases reduction declared at the Climate Summit in Paris as 75–70% of 1990.

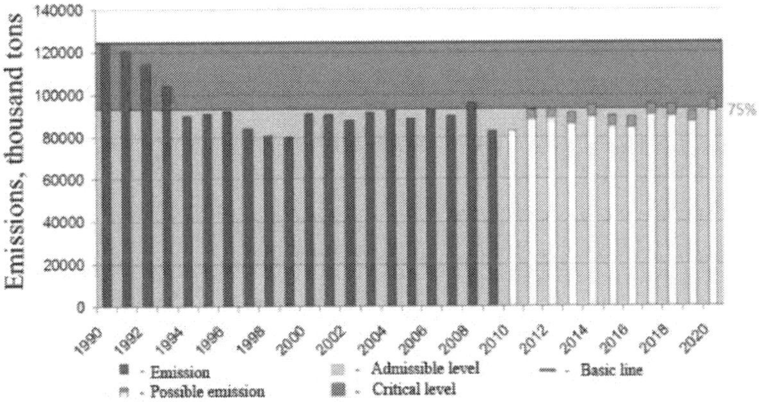

Fig. 1. Emissions of greenhouse gases coordinated with the level of the RF obligations up to 2020.

Taking into account the positive trend in greenhouse gas emissions approximately by 0.67% per year and the dynamics of Sverdlovskaya Oblast economic development, it creates prerequisites for further intense growth of greenhouse gases emissions. In the nearest perspective for 2017–2020 we exceed the level stated by the President by 5–9%.

Figure 1 shows that in 2017 and 2018 the anticipated increase will be about 2%, and by 2020 the emissions of greenhouse gases can exceed the level of obligations accepted in Sverdlovskaya Oblast by 4–5% [14].

Analysis showed the decrease of emissions of greenhouse gases is the responsibility of the sphere of electric energy production (decrease of specific fuel consumption per unit of supplied electricity) and metallurgy (increase of power efficiency).

Nowadays the problems with emission of greenhouse gases are not taken seriously by the government, science, business and citizens, though the fulfillment by enterprises in Sverdlovskaya Oblast requires the development of low carbon strategy in the region.

Conclusion

The scheme of transition to a new model of green economy is presented in the article as well as its acceleration by activation the creative and intellectual potential of human. It will allow the decrease of impact of pollutants and greenhouse gases, climate change, as well as the risks of natural and manmade disasters rise and decrease their consequences (Figure 2).

The problem of global climate change should be considered fully because global climate change, natural and manmade disasters are closely related to energy, economic, climate, and environmental problems in the "nature-society-human" system. Not only scientists and officials but ordinary people should understand them and be closely familiar with them. With that end in view the scheme should involve human capital that is able to start low-carbon power-effective mechanism in maim sectors of Russian economy. The idea itself and transition from resource to green economy is a large-scale innovation covering all the sectors of the economy, and it begins from energy efficiency.

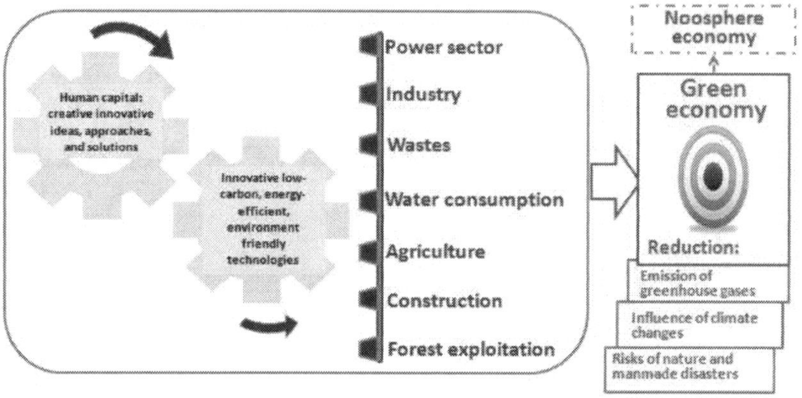

Fig. 2. Scheme of initiation of the transition to green (smart) economy model.

Figure 3 shows the total economic effect of:

•reducing consumption of fuel-energy resources;

•prevented environmental damage by reducing the amounts of pollutants and energy resources consumption;

•reducing the carbon footprint by decreasing emission of CO_2-equivalent;

•reducing the risk of natural and manmade disasters due to reducing emissions of greenhouse gases.

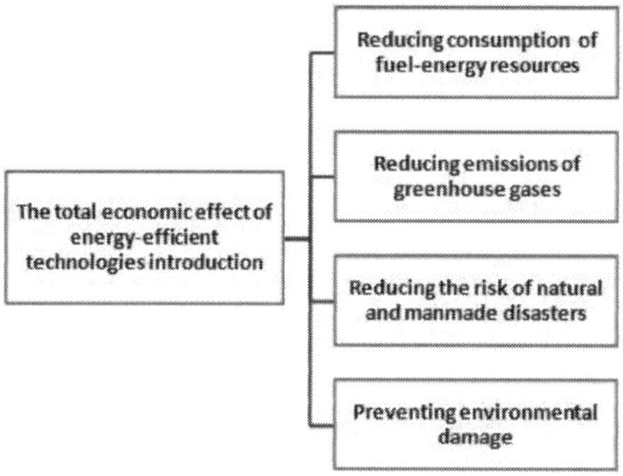

Fig. 3. Total economic effect of introduction of innovative power-efficient low-carbon technologies.

That is innovative solutions of energy saving and low-carbon strategy is the key trends of the Russian economy for a long period of time.

References

1. Russian population: size, dynamics, statistics, http://www.statdata.ru/russia

2. Renewables 2015 – Global Status Report, http://www.ren21.net/wp-content/uploads/2015/07/REN12-GSR2015_Onlinebook_low1.pdf

3. F.F. Kurbanova, I.M. Bilalova, *Information package of the VIII International Student Electronic Scientific Conference "Student Scientific Forum"* (2016)

4. In Moscow's Expo Center a prototype of "a smart city" was built, http://expert.ru/2015/11/12/v-moskovskom-ekspotsentre-postroili-prototip-umnogo-goroda
5. City of Vancouver: the official website of Vancouver City Administration, http://vancouver.ca/green-vancouver/targets-and-priority-actions.aspx
6. ECO2 Cities: Ecological Cities as Economic Cities, http://www.esmap.org/sites/esmap.org/files/CS_Stockholm.pdf
7. Li Kuan Yu, *The history of Singapore. From «the third world» to the first one»* (2005)
8. O.L. Kuznetsov, P.G. Kuznetsov, B.E. Bol'shakov, *The system nature – society – people: sustainable development* (Dubna, 2000)
9. V.I. Vernadskiy, *Biosphere and Noosphere* (2004)
10. A.I. Bedriy. Creative economy, **5** (**77**), 16–21 (2013)
11. How to guide: Low-emission Development Strategies and Nationally Appropriate Mitigation Actions: Eastern Europe and CIS. http://www.undp.org/content/dam/undp/library/Environment%20and%20Energy/Climate%20Change/Mitigation/Low-emission%20Development%20Strategies%20and%20Nationallv%20Appropriate%20Mitigation%20
12. Renewable energy. Medium-term market report 2015. Executive summary. https://www.iea.org/Textbase/npsum/MTrenew2015sum.pdf
13. Climate changes – business won't find it funny. http://www.rusecoiinion.ru/businessl40316
14. V.P. Anufriev, S.A. Yazykov, A.A. Panchenko, *The Materials of the 5th International Theoretical and Practical Conference "Efficient and Quality Electric Energy Supply and Utilization"* (Ekaterinburg, 2016)

Development and metabolism of the city of Khartoum (Republic of Sudan): spatial designing of the coastal territory of the Blue and White Nile

Ilhomjeon Shakurov[1a], *Ahmed* Elamin[1], and *Mohammed* Zebilila[1]

[1]Moscow State University of Civil Engineering, 129337, Moscow, Russia

Abstract. The given article reveals the development of the metabolism of the city of Khartoum, the problems of influence on the microclimate improvement elements built up coastal territory of the Blue and White Nile in Khartoum. The study shows the microclimate formation factors of Khartoum, gives an assessment of the modern state of the microclimate of the coastal zone and its changes. The article indicates that, depending on the nature of use and planning organization of coastal territories microclimatic influence of water areas can propagate over appreciable distances to the city or limitation depth in a narrow strip along the coast. In urban areas the air temperature is considerably higher and the humidity is lower than in the coastal area especially in summer, this is due to differences in physical properties of water and land. Water has a higher heat capacity, high reflectivity, the wind speed increases and increasing the humidity of air masses passing over it. Thus, in order to improve microclimatic and hygienic conditions of coastal territories in Khartoum there is the need to use the favorable impact of the Blue and White Nile.

Sudan, a country of deserts and savannah, but this general definition hides a wide variety of environmental conditions. It is not due to the so much differences in topography (as most of the country is slightly dissected plateau), but the difference in the latitudinal position, which determines the climatic conditions. The territory of the country is located within an ancient African platform. Change of the width zones – from the "classical" desert in the north to rainforests in the south is well expressed. The surface of Sudan – almost a plain raised on height of 500–1000 m above sea level. Its heights increase to the West, the East and the South.

The largest city of Khartoum – the capital of the Republic of Sudan, the state in the northeast of Africa. Khartoum is located in the central part of Sudan, located in a place where the White and Blue Nile are merged, and is flowing as the Nile River, which is flowing through Egypt and flows into the Mediterranean Sea. The city is located on the elongated promontory on the left bank of the Blue Nile at an altitude of 400 m above sea level. The confluence is of the form of an elephant's trunk, (al-Khartûm – an elephant's trunk in Arabic). Hence the name of the city [1].

[a] Corresponding author: shukurov2007@yandex.ru

The climate of Khartoum is characterized by:

- Exhausting warm summer climate;
- Large volatility and low relative humidity during the hot period;
- A small annual sum of precipitation;
- Slight wind speed;
- Availability of breeze phenomena with a daily change in direction of coastal areas of the Blue and White Nile.

With a research objective of finding the influence of the Blue and White Nile on coastal strips we carried out the analysis of the design and experience of construction of coastal settlement, and field observation in this territory [2].

An analysis of existing work and experience in the design of the coastal strip in Khartoum allows us to conclude that at present there is no generally agreed evidence-based approach for accounting the impact of the Blue and White Nile in the urban metabolism and microclimate of the city of Khartoum.

In several cases, a differentiated approach is not practiced to the planning and settlement of separate urban areas, which are located in complicated geographical conditions and are characterized by various levels of beautification.

In carrying out field observations for the changes in temperature and humidity Assman's psychrometer and M-22 s and M-16 s recorders were used. Speed and the direction of the wind were measured by M.s-13 cup anemometer and a directional sign. Before measurement, calibration of devices was carried out. The general weather conditions near Khartoum were taken from the meteorological station "Airport".

In observing the temperature, humidity and wind mode there was a simultaneous revelation of the influence of green plantings and types, underlying surfaces with different radiation characteristics in coastal and remote areas, located in the settlement, as well as out of it [3].

Observations have shown that during the day the temperature of sun-lit area of the surface of the sandy soil, lawn is on the average lower by 15–20 °C, than the temperature of the surface of loamy soil and asphalt, and the maximum difference is 45 °C. The maximum surface temperature observed was from the period from 8 hours to 19 hours.

It is known that the underlying surface heats adjacent layers of air. It is therefore expected that air temperature in all cases has to be lower than temperature of the underlying surface. However, it was only appropriate for the lit surfaces. Air temperature over sand was 4–6 °C lower in the morning and in the evening, and the maximum difference of 42.4 °C was observed at 17 hours. Over the lawn, this difference of 4 °C was constant until 16 o'clock, and subsequent air temperature exceeded lawn surface temperature by 3–5 °C. The air temperature over the lawn and sand is below that over asphalt, on average by 4.8 °C. Sometimes this difference reached 5 °C. The air temperature over the sand and lawn were characterized by small amplitudes of fluctuations during the period of observation as seen in similar data over asphalt pavements [4].

Air temperature over a lawn and the soil under tree crowns (in shade) is always below, that over asphalt pavements, and the difference in temperature reaches 48 °C. Distinction of air temperatures over the lit and shaded surfaces of the lawn and sand was about 8.5–11.5 °C. Under crowns, relative humidity of air is by 15–20% more than similar lit underlying surfaces.

Process of change of air temperature by green plantings was seen to be the opposite process of the change of air temperature of the underlying surfaces. Under tree crowns, air temperature was higher than soil surface temperature, and the difference reached 11–14 °C. This phenomenon is explained by advection of air masses from more heated adjacent surfaces.

In the zone of influence of the Blue and White Nile green plantings are capable of preserving such influence on a microclimate settlement. So, with a distance from the rivers, in a zone of green plantings 250 m wide the change of air temperature was 2–2.8 °C, and humidity of air was only 12–14%. Whereas in territories that were not planted with trees and shrubs the corresponding differences reached 1–3.5 °C, and humidity of 4–10%. Wind speed in a green zone far from the rivers decreased by two and a half times, than in the coastal territory which was not planted with trees and shrubs in similar conditions. Paths between green plantings promote the direct flow of air from Blue and White Nile to the settlement [5].

In studying the influence of the Blue and White Nile on the urban metabolism and microclimate mode of the city of Khartoum, we conducted field observations. The field observations carried out determined that the change of the microclimate parameters of the settlement was due to the influence of the Blue and White Nile, which is noticed in areas of 2 200 m wide. With increasing distance from the river predominant impact on microclimate areas are on buildings, artificial pavement, beautification and the like elements. The degree of influence of the river changes throughout the day. Thus during morning hours air temperature difference between the water and land surface layers is 3–7 ° C, and 6–9 °C in the daytime. Air humidity changes between 20–24%, wind speed from 0 to 2.5 m/s. Changes in temperature and humidity and wind regime in the development, which is formed under the influence of the river is shown in Figure 1.

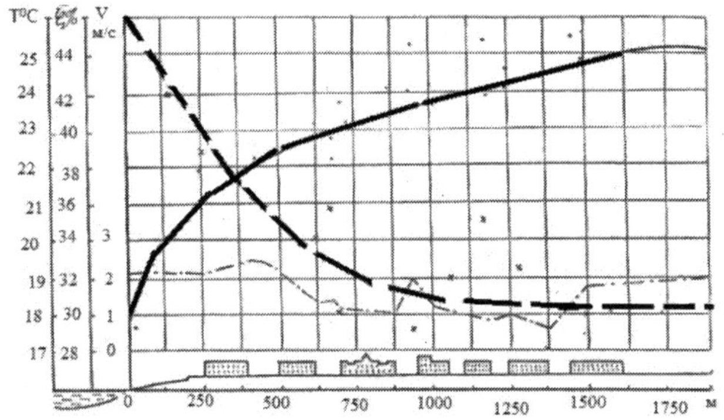

Fig.1. Change of microclimatic parameters of a microclimate in building settlement of a coastal zone.

Irregularity and different nature of microclimatic characteristic plots of urban environment also determines the various planning decisions of housing areas, especially in the Blue and White Nile.

Based on the results of the field observations changing in air humidity in the settlement $\Delta\overline{\varphi}$ is determined by the formula:

$$\Delta\overline{\varphi} = \left(\overline{\varphi}_B - \overline{\varphi}_c\right) \times \left(1 - m_B \ln^2 L_B\right).n.d, \quad (1)$$

where $\overline{\varphi}_B$ – Air humidity from the water's edge at a height 1.5 m from the ground level, %;
$\overline{\varphi}_c$ – Air humidity outside the zone of influence of the river to the settlement at a height of 1.5 m from the ground, %;
m_B – Coefficient depending on the size of the river in the direction of the wind, the ratio of the turbulence coefficient to the wind speed equal to 0.02;
L_B – Width of the zone of the river's influence to air humidity, m;
n – The coefficient considering the change in speed of wind depending on the height of buildings and the locality (territory roughness);

d – The aerodynamic coefficient for vertical surfaces, taken as 1.1 for windward side and 0.85 for leeward side.

Air streams generally come from Blue and White Nile. The wind pattern of building settlement depends on the direction of the dominating wind and the planning scheme of streets. Thus, despite variety in the directions of a wind, there exists a well-defined flow path of wind. So, for example, for winds coming from the river, the main air streams are directed along streets, perpendicular to the riverbed of Blue and White Nile [6].

On the entrance to building settlement, under the influence of compressed wind stream the relative speed of the wind is 0.2–0.4 higher than speed in undeveloped territories. In building settlement wind streams are sent along streets and through spaces between buildings.

The disorderly movement of wind streams is noted at intersections, on streets, on undeveloped sites, between buildings and trees. Between buildings of width 0.6–0.8 H wide (H – the accepted building height) the speed of the wind increases in gaps by 0.2–0.4 of the speed in undeveloped territory. The air stream is divided into separate streams, part of which comes to quarters of the cross street.

At a deviation from the main direction of a wind at an angle less than 45 ° the direction of the movement of streams on streets remains the same, as well as the direction of the main wind taking the direction of the longitudinal streets.

At a wind direction of an angle closer to 45 ° to the front of the building settlement the main air streams go in the directions seeking to meet with the main wind direction, i.e. as if in steps formed by sites of mixed and crossed streets [7].

Thus, the wind directed from the river is capable of increasing the extent of influence of Blue and White Nile on the temperature, humidity and wind pattern of the building settlement of coastal zone. For wind streams of opposite direction, the impact of the rivers on the microclimate of the building settlement is less significant.

Based on the conducted field research, a microclimatic zoning for the territories of the city was carried out. The following are characteristic zones developed for the city:

The first area, which is directly adjoining Blue, White Nile and beginning from the water edge where, parameters of a microclimate have values close to values of surface layer of air. So, changes of temperature, humidity and wind pattern is: $\Delta T = 4–9$ °C, $\Delta\overline{\varphi} = 12–24\%$, To = 1.2–2.5.

The second area – a buffer zone along building settlement of the coastal territory, which the settlement has an impact on. The width of this zone is about 160–250 m and where changes of parameters of a microclimate respectively is $\Delta T = 2–7$ °C, $\Delta\overline{\varphi} = 6–18\%$, To = 0.8–1.8.

The third area – under direct influence of the river on the formation and change of a microclimate of building settlement 800 m wide. In this zone change of parameters of the microclimate respectively is: $\Delta T = 0.8–4.5$ °C, $\Delta\overline{\varphi} = 8–13\%$, To = 0.7–1.5.

The fourth area – periodic influence of the river depending on the direction of the wind and the breeze phenomena in the coastal zone. In this zone the influence of the "city" on formation of the temperature, humidity and wind pattern is more noted: $\Delta T = 0.4–1.5$ °C, $\Delta\overline{\varphi} = 4–8\%$, To = 0.3–0.9.

The fifth area – with a small density of buildings with considerable open spaces and the areas of artificial coverings, where influence of the river affects to a lesser extent.

The sixth area with building blocks, creating closed limited spaces with the microclimate, which is formed under the influence of beautification elements. Changes of the temperature, humidity and wind conditions have small values and influences ($\Delta T = 0.5–1$ °C are explained by a conservative nature and resilience to external influence; $\Delta\overline{\varphi} = 6–12\%$; To = 0.2–0.6).

According to the degree of influence on the microclimate of the building settlement by the river four zones of the territory of Khartoum are identified:

a) Lower terrace of a coastal slope;
b) Zone between a brow of a coastal slope and the front of building settlement;
c) The built-up territories experiencing continuous microclimatic influence of the river;
d) The built-up territories coming under the incidental (unstable) influence of water bodies.

Thus, the analysis of results of the observation allows to draw similar conclusion, that the underlying surfaces possessing property to reduce amplitudes of fluctuation of temperature and humidity of air (lawns, bushes, green plantings) promote improvement of the microclimate of the territory and increase in the zone of influence of Blue and White Nile.

A small area of trees and plants contribute to the overall reduction of air temperature and reduction in humidity, comparable to filters, allowing in air of the surrounding areas and cooling it. Thus, by appropriate landscaping it is possible to regulate micro-climatic conditions of the building settlement.

Planting of trees and plants in the zone of influence of Blue and White Nile promotes the positive impact on the temperature, humidity and wind condition of coastal territories of Blue and White Nile.

References

1. H. Adam, *Features planning regular reconstruction of residential one-storey and elemental chaotic building: E.g. Khartoum, Sudan* (2002)
2. A. Giyasov, O. Sokolskaya, O. Al Fakri, UP, **6**, 58–62 (2012)
3. *Guidelines for an assessment and regulation of the wind regime of residential development* (1986)
4. S. Chistjakova, C.P. house, **3**, 20–28 (1965)
5. I. Shukurov, *Problems of the warm and wind mode of the dwelling of Iraq* (2008)
6. I. Shukurov, *L.I.&M*, **3**, 43–46 (2013)
7. I. Shukurov, *H.C*, **2**, 20–21 (2005).

E3S Web of Conferences 6, 01011 (2016) DOI: 10.1051/e3sconf/20160601011
ICSC 2016

Role of the University in city agglomerations sustainable development

Sergey Kortov[1], *Alexandra* Terlyga[1a], and *Nadezhda* Terlyga[1]

[1]Ural Federal University named after the first President of Russia B.N. Yeltsin, 620002, Yekaterinburg, Russia

Abstract. It is shown that existing models of sustainable cities and entrepreneurial universities development do not take mutual influence into consideration. In this work there are offered approaches to the models of interconnected cities and universities on the basis of the principle «won-won», principle of scale accountancy and dynamics of university activity results influence on a sustainable city development is validated, systems of conjugated markers and indicators are elaborated.

1 Introduction

Chain of crisis phenomena shaking global economics during the last 15 years, accelerated development of technics and technologies, change of social processes tell about the growth of economic formation which is more and more often called as postindustrial society. The brightest characteristic of a new civilization development period is given by E. Toffler [1], who characterized it as "The Third Wave": "The Third Wave brings with it a genuinely new way of life based on diversified, renewable energy sources; on methods of production that make most factory assembly lines obsolete; on a novel institution that might be called the "electronic cottage";, on and on radically changed schools and corporations of the future». Transition to a new formation goes through acute aggravation of demographic ecological, technological, social cultural problems of society. One of the answers to these challenges is the idea of «sustainable development», which was first stated in 1987 in a form of the concept in UNO Committee report on environment and development [2] and it is performed as 27 principles of sustainable development in 1992 in UNO Declaration on environment and development [3].

At its core the concept of sustainable development has an anthropocentric character, defining actions in three main directions: ecological, social and economic. In Russia the concept of sustainable development was accepted in the form of Decree by the President of the RF «Main statements of RF strategies on environmental protection and sustainable development» [4]. Principles of sustainable development are successfully applied to elaborate strategic documents for the countries, separate territories and cities [5–7], to implement the development programs management tools are intensively worked out. It includes as a rule structured lists of indicators for strategic and operating planning and responsivity [8–10].

[a] Corresponding author: sterlyga@gmail.com

In conditions of postindustrial society special role is played by knowledge as the source of unique competitive advantages for economical agents, territories and states [11]. In this case the role of universities as generators of knowledge, human capital assets and key elements of national innovation systems rises sharply. Transformation of universities from «cathedral of science» into a significant factor of competitive struggle on a global level is reflected in the concept of entrepreneurial university [12, 13].

From the viewpoint of economic model the University of Entrepreneurial Type is a source of competitive growth, incomes and taxes flow, but not a center of state and business expenditures for science and education. University presence in the area must be a driver of its social economic development. In the work [14] there are 4 directions of transformation offered for entrepreneurial university. They are connected with the availability of competitive development strategies, entrepreneurial type of university management, formation of students and teachers' entrepreneurial culture, income sources diversification. As a matter of fact we are talking about presence of entrepreneurial university strategy connected with commercialization of its key competences in education, scientific research, innovation products and services, providing services not only from the vendor but also from the integrator of large-scale development projects for business, state and society. This task is stated in the definition of «the third mission», in the framework of which universities become «active participants of economic and cultural processes development; convert to organizations closely connected with industry and society as a whole» [15].

Analyzing the systems of indicators which characterize universities entrepreneurial potential, performed in a series of works [16, 17] shows that they are focused mainly on revealing inner factors of entrepreneurial medium development though a detailed analysis of university activity results influence on economics and society is absent.

Historical way of universities development shows that they are predominantly localized in city environment. Nevertheless in sustainable cities development concepts and strategies universities are positioned in their classical role only as a source to get a higher education for population. On the other hand entrepreneurial universities are basically aimed to achieve competitiveness in global markets and little attention is paid to the tasks of local territories and cities development. To some extent university influence on social-economic territory development happens automatically proceeding from natural «congruence» of the university as a part of the city. Although efficiency growth of entrepreneurial university as the city development driver requires transition to strategic management, where elaboration of city and university sustainable development inclusive model is needed. The aim of the present work is formation of approaches to create associated indicative models of city and entrepreneurial university sustainable development.

2 Analysis of approaches for measurement of city and entrepreneurial university sustainable development

To assess sustainable cities development mainly an indicative approach is used. It is based either on indicators offered by committee on sustainable UNO development either on authorial concept of social-economic development strategies vendors (see, for example, [7]).

In RF the system of sustainable development indicators is elaborated [18], it includes 4 groups, 18 subgroups and 132 indicators:

- social indicators (antipoverty activity, demography, health protection, education, transportation);
- economic indicators (revenues, consumption, finance);

- ecological indicators (water, land and other resources, waste, atmosphere);
- institutes (presence of sustainable development strategy and population involvement in its implementation).

For rating score of cities sustainable development indicators systems made by ranking agencies are used. In McKinsey model [19] 5 directions, 18 factors and 18 indicators are performed:

- main needs of the population (water supply, accommodation, health protection, education;
- resources utilization efficiency (power, water, waste, industrial load);
- environment cleanness (pollution, purification, waste recycling);
- city infrastructure (transportation, population density, planting, buildings power efficiency;
- focus on sustainable future development (investments to environmental protection and ecologically clean work places).

In the rating of sustainable cities development, worked out by the company SGM [20] there are 4 groups which include 31 indicators (table 1):

- demography and population;
- social infrastructure;
- city infrastructure;
- economic development;
- ecology.

Analysis of sustainable development indicators system shows that universities are present in capacity of social subsystem element only in the model of SGM company (number of higher education establishment students for 10000 inhabitants). In all other models attention is focused on secondary education availability. At this rate the role of universities as drivers of sustainable cities development in majority of existing models is not reflected, although expectations connected with them are present in strategies of the Russian Federation regions social-economic development (see, for example [21]).

Let us consider systems of entrepreneurial universities development indicators. Most of the entrepreneurial universities models are founded on 5 principles by B. Clark. According to this work [22] in the frame of entrepreneurial universities research in Europe the model including 7 branches was used. These branches are:

- leadership and management;
- organizational potential, people, inducements;
- development of entrepreneurship in teaching and knowledge mastering;
- development possibilities for entrepreneurs;
- collaboration between university and business;
- international communications;
- level of university influence on external environment development.

In the work [16] in the frame of Russian universities research entrepreneurial potential modified model was used [22], which consists 6 characteristics groups.

Strategy:

- understanding of entrepreneurship development as a strategic aim of the higher education establishment;
- entrepreneurship activity growth support;
- inducements and awards for the teachers and researchers who actively support students' entrepreneurship.

Finance resources:

- long-term financial support of student enterprises as an approved part of university budget;

- self-sufficiency of the university in the sphere of inner entrepreneurship support.

Human resources

- higher-education teaching personnel career enrolment and development with consideration of stance on entrepreneurship, presence of corresponding behavior and experience and also entrepreneurship activity support;
- development of human resources necessary for initial support of entrepreneurship projects lead by students and employees.

Entrepreneurship and start ups support

- presence of special entrepreneurship support infrastructure, conditions for business-hatching in campus or help in providing access to external objects;
- close collaboration with development institutes;
- integration of entrepreneurship education and support processes;
- including the university to the channels of access to private financing;
- mentoring from the viewpoint of teachers and entrepreneurs;
- integration of entrepreneurship support processes inside and outside.

Entrepreneurship education

- including entrepreneurship education in curriculum programs;
- wide offering of different forms of fast and effective education;
- creative methods of teaching, students and post-graduate students requirement accountability;
- dividing offers for burgeoning entrepreneurs and owners of growing business;
- students and graduates business support as a key component of education;
- integration into entrepreneurship training the results and research of its condition and growth.

Assessment

- regular inventory counts and entrepreneurship activity efficiency audit;
- formalized assessment of entrepreneurship activity, which includes long term monitoring of graduates entrepreneurship activity development.

As it is seen from the described approaches to the entrepreneurial university assessment, the main focus is on the entrepreneurship training and students and teachers entrepreneurial activity support which is expressed in creation and development of small enterprises on the basis of university elaborations. In a wider understanding on entrepreneurial strategy, offered in work [14], it is offered to assess entrepreneurial behavior as strive for key competences commercialization not only in the innovations sphere but also in education and scientific research.

Analysis of main assessment models of university entrepreneurial strategy shows that university is observed as an open system collaborating with external environment via results of its activity and influencing on it in a significant way. During the analysis of external environment development in the work [22] it is assessed how university entrepreneurial activity influenced on its surrounding and graduates, in the work [16] it is recommended to lead regular monitoring of graduates entrepreneurial activity. While evaluating the universities entrepreneurial potential the main is an expert method realized via target groups questionnaire with further results processing with the help of grade-rating assessment. In Russia an indicative method of entrepreneurial potential assessment is almost impractical, because entrepreneurial universities activity is weakly reflected in governmental system of their efficiency monitoring indicators. There is an interesting entrepreneurial and innovation universities activity level rate assessment model by Interfax agency [24], where 7 indicative markers are included:

1. Innovation entrepreneurship development level in higher educational establishment.
2. Volumes of patents network (national and international), supported by higher educational establishment by the end of 2014.

3. Participation of higher educational establishment in technological platforms development, in Programs of high-technological companies' innovation development.

4. Participation of higher education establishment in technological entrepreneurship development programs.

5. Educational programs (mainstream and supplementary education) performed by higher educational establishments in the sphere of entrepreneurship.

6. Volume of economic-contractual works in budget of higher educational establishment in 2014.

7. Participation of academic staff and learners of the higher educational establishment in objects of innovation infrastructure.

Ranking is performed via questionnaires (expert assessments) and application of state statistics data (indicative rates).

As it is seen from provided analysis, entrepreneurial universities are aimed to measurement of it influence on surrounding although development of the city and the territory as a life environment are not emphasized as a priority strategy in entrepreneurial university indicators system. Unfortunately, neither cities nor universities witness each other as natural partners who create competitive advantages for each other in case of elaboration and realization of congruent strategies of sustainable development. Working out a mutually connected system of university and city stable development is an important task, which solution will allow assessing the degree of mutual influence and compound documents of strategic character defining forms, mechanisms and results of such collaboration.

3 Opportunities for creating of stable city and entrepreneurial university development congruent indicative models

In this work the analysis of activity results congruency of entrepreneurial university on main cities stable development indicators was performed. As a basis the model offered in work [20] was taken, it is performed in table 1. The authors led a qualitative expert evaluation of university activity influence on a current indicators level as well as on their alterations rates. Assessment was made on the example of Ural Federal University (UrFU has more than 30 thousand students and 7 thousand employees, 9th place among the universities of Russia according to the Interfax ranking «Innovations and entrepreneurship» in 2014) on corresponding indicators of Ekaterinburg. For assessment the following scale was used (0 – influence is absent, 1 – weak influence, 2 – average influence, 3 – strong influence). Near indicators in table 1 explanation on UrFU activity results indicator influence are given. Average values of UrFU on current indicators condition and rates of their alteration comprise 1.29/1.51. The greatest impact is made by university on education system development and city economics. From table 1 it is seen that the following list of indicators reflects not all the opportunities of university influence on the city environment. It is suggested to update the model with 3 new groups of indicators:

- development of small and medium businesses (share of small and medium enterprises (SME), volume of products issued by SME on the unit of economically active population (EAP));
- innovation activity of enterprises (share of innovation – active enterprises, share of innovation production in manufacturing volume, level of new technologies application in goods and services production);
- youth policy (relative share of employed youth, level of youth access to entrepreneurship, level of youth activity in healthy way of life and ecological initiatives);

Assessment of the offered indicators influence in the system of table 1 increases correlation of university activity results influence on stable city development up to 1,53/1,90. Further indicators system development for the cities in the context of collaboration with the university will allow shaping congruent pictures of competitiveness, by the reference to a principle «won-won».

Table 1. Entrepreneurial university influence on sustainable city development indicators (on example of UrFU and city of Ekaterinburg)

	Influence on present value/on themes of alteration	**List of RF cities stable development indicators**	
entrepreneurial potential of the university	1/1 (birth rate of students and employees)	Demography	Coefficient os natural increase, %
	2/3 (acceptance of foreign and nonresident students)		Coefficient of migration increase, %
	2/3 (middle-age decrease)		Demographic load, %
	1/1 (presence of policlinic, UrFU)	Medicine	Number of doctors per 10000 of population, %
	1/1 (presence of policlinic, UrFU)		Number of mid-level health professionals per 10000 of population, %
	0/0		Number of beds at in-patient facilities per 10000 of population, %
	1/1 (presence of policlinic, UrFU)		Capacity of outpatient departments per 10000 of population,
	1/1 (availability of kindergarden by the university)	Education	Number in need of attending pre-school institutions as regards places in them, %
	1/1 (presence of secondary specialized college at the university)		Number of secondary specialized college students per 10000 of population, %
	4/4 (more than 30% of students of Ekaterinburg)		Number of students per 10000 of population, %
	1/0 (presence of safety protection system at UrFU)	Safety	Number of registered crimes per 1000 of population
	1/2 (dormitories building)	Housing facilities	Housing supply per head, m²/person
	1/2 (building of new dormitories and corporate housing)		Share of families in the improved housing queue, %
	1/2 (building of new dormitories and corporate housing)		Renewal of housing facilities, %
	1/0 (wreckage of old dormitories)		Share of rundown and dilapidated housing, %
	2/2 (buiding of comfortable dormitories)		Housing supply by systems of heating, water facilities
	2/2 (presence of own heat source)	Public utilities	Average yearly heat supply source output, thous. gkal
	2/2 (spreading of campus all over the city)	Transportation	Number of trips per 1 person on local transportation yearly, thousand times
	2/3 (85 small enterprises, implementation of UrFU	Production, investments	Industrial production per 1 representative of economically active population (EAP),

	elaborations in enterprises		thous. rub./ person.
	1/2 (implementation of UrFU elaborations in enterprises)		Investments in main capital per 1 representative of EAP, thous. rub./person
	0/0		Relative share of loss-makers in a whole circle of organizations, %
	0/0 (graduates registration statistics in labour market is absent)	Labor market	Registered unemployment, %
	0/0		Market diversity (Herfindahl-Hirschman index)
	2/3 (UrFU is one of the biggest taxpayers)	Budget	Share of own incomes in city budget incomes, %
	2/3 (average salary level of UrFU is 2 times higher than an average one in the city)	Income and expenditures of population	Reference of organizations employees salary to minimum living wage, %
	2/2 (students and employees incomings and purchasing power growth)		Turover of retail per 10000 of population, %
	2/2 (students and employees incomings and purchasing power growth)		Turnover of public catering per 10000 of population, %
	1/2 (UrFU elaborations introduction in enterprises and residential buildings)	Production ecology	Water consumption per unit of industrial production, m^3/thous. rub.
	1/2(UrFU elaborations introduction in enterprises)	Environment condition	Pollution discharge per 1 km^2 of city area
	1/1 (concentration of students in campus)		Population density, person/km^2
	1/1 (boiler house presence)		Integrated air impurity index (AII)

In the models of entrepreneurial university, in the framework of the models considered in the article, there is a unit of indicators connected with the assessment of university influence on economics and society. It is offered to supplement them with the following groups and elements:

- scale of university influence on social, economical and ecological development of the city in measuring employment, goods and services production volume, taxation base growth, salary growth, demographic indicators improvement;
- level of impact on the indicators of innovation development, connected with the transfer of new technologies in the sphere of industry and services, development of high technological clusters, increase of innovation products output and export, increase of economics research intensity, development of labour productivity, intellectual property objects application;
- level of influence on entrepreneurial climate of the city is defined by number of small and medium enterprises, created with the help of the university, development of technical parks and business-incubators, increase of entrepreneurial culture and education level;
- level of impact on population social and ecological activity is connected with active youth policy of the university, influence on public organizations work, expert community, leading of cultural and ecological events.

Expert assessment (according to 5 grade scale) of UrFU entrepreneurial potential in 2015 in the framework of the work model [16] with consideration of offered influence indicators is shown in picture 1 on the basis of three target expert groups cancass – representatives of the university, business and authorities.

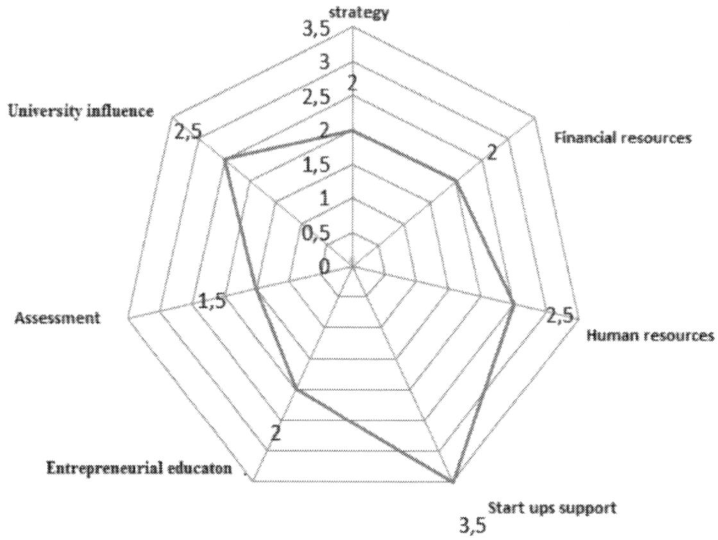

Fig. 1. UrFU entrepreneurial potential assessment.

Model of sustainable city development with new indicator groups offered in the article, and model of entrepreneurial university with the unit of university indicators of impact on economics and society are congruent, i.e. accounting scale and dynamics of mutual influence. Substantial disadvantage is a different nature of indicators in these models: indicative for cities and expert for universities, which requires working out indicative markers of entrepreneurial university potential development.

4 Conclusions

Existing models of sustainable cities development and entrepreneurial universities do not take their mutual influence into consideration and do not allow to work out inclusive strategies of competitive advantages shaping taking into account mutual opportunities.

As regards growing universities role in cities development, the work offers approaches to creation of congruent cities and universities models on the basis of the principle «won-won», the principle of considering scale and dynamics of university activity influence on sustainable city development, systems of conjugate exponents and indicators were worked out.

For further application of this approach it is necessary to develop the system of quantitative indicators, reflecting entrepreneurial university growth, methods of their accounting and analysis, performance of cities indicators research, those cities where entrepreneurial universities are present.

Acknowledgements

This research was supported by Act 211 Government of the Russian Federation, agreement № 02.A03.21.0006

References

1. Alvin Toffler, *The Third Wave (*2004)
2. Our mutual future. Report of International Committee on environment and development (1989)
3. Rio de Janeiro declaration on environment and development (1992), http://www.un.org/ru/documents/decl_conv/declarations/riodecl.shtml
4. Decree of RF President № 236 from 04.02.94, http://www.kremlin.ru/acts/bank/5422
5. LEIPZIG CHARTER on Sustainable European Cities (2007), http://ec.europa.eu/regional_policy/archive/themes/urban/leipzig_charter.pdf
6. Reference Framework for European Sustainable Cities, www.slideshare.net/RFSC
7. Concept of sustainable development of Sochi up the year 2010, http://www.sochiadm.ru/upload/iblock/aaa/aaa984be6c24cb8caecc072fb8f3dbc4.pdf
8. Indicators of Sustainable Development, UN Department for Policy Coordination and Sustainable Development, December (1994)
9. Human Development Report 2003. Millennium Development Goals: A compact among nations to end human poverty (2003), http://hdr.undp.org/sites/default/files/reports/264/hdr_2003_en_complete.pdf
10. S.N. Bobyileva, P.A. Makeenko, *Indicators of Sustainable Development of Russia (ecological-economical aspects*) (2001)
11. J. Houghton, P. Sheehan, *A Primer on the Knowledge Economy* (2000).
12. B. Clark, Journal of the Programme on Institutional Management in Higher Education, **13** (**2**), 9–24 (2001)
13. G.N. Konstantinov, S.R. Philonovich, Education issues, **1**, 49–62 (2007)
14. S.V. Kortov, D.B. Shulgin, N.G. Terlyga, Innovations, **8 (190)**, 42–50 (2014)
15. H. Etzkowitz, A. Webster, C. Gebhardt, B.R.C. Terra, Research Policy, **29** (**2**), 313–330 (2000)
16. D. Williams, A. Kluev, Industry and Higher Education, **28(4),** 271–280 (2014)
17. A. Kuzmin, G. Podolniy, Designing, monitoring and assessment, **2,** 2–18 (2015)
18. *New paradygm of Russia development in XXI century. Complex research of sustainable development problems: ideas and results* (2000)
19. New models of cities development. View McKinsey & Company. http://gtmarket.ru/laboratory/expertize/5974
20. Rating of sustainable cities development in RF during the year 2012 http://agencysgm.com/upload/iblock/66b/reyting_2012.pdf
21. Strategy of social-economic development in Sverdlovk region. http://smb.gov66.ru/content/legislation/regionalleg/act_adm_reg/m,65,1320
22. B. Clark, Higher Education Policy, **17**, 355–370 (2004)
23. OECD. Universities, innovation and entrepreneurship. Criteria and examples of good practice. http://www.oecd.org/regional/leed/43201452.pdf
24. National rating of the Russian universities by Interfax agency, http://www.univer-rating.ru/txt.asp?rbr=30

E3S Web of Conferences 6, 01012 (2016) DOI: 10.1051/e3sconf/20160601012
ICSC 2016

Regulatory support of sustainable intermodal transport transit hub system development

Denis Vlasov[1, 2a]

[1]Moscow State University of Civil Engineering, 129337, Moscow, Russia
[2]Genplan Institute of Moscow, 125047, Moscow, Russia

Abstract. The national urban development policy gives more and more attention to the most essential element of the City Transport Structure Planning – Intermodal Transport Transit Hubs. There is a significant number of ITTU project plans being developed in Moscow, Moscow region and different regions of the Russian Federation. While the regulatory framework for the development of documentation for ITTU territory planning is almost absent. This publication presents the author's point of view on the structure and composition of regulatory documentation for ITTU urban design. This article reports the results of the author's original scientific research on the development of ITTU systems.

1 Introduction

The transport infrastructure development is the priority direction of socio-economic urban planning in the Russian Federation. Modern, developed, efficient transport infrastructure is a key factor that ensures sustainable development of the whole country as well as of every region and municipality.

Recently, national urban planning policy pays much more attention to the development of various intermodal transport transit hubs (ITTH). A transport hub is the element of the city communication network that enables interaction between different modes of transport infrastructure as well as the interaction between city transport and serviced territory.

2 Justification of regulatory documentation development

Actually, ITTH is a miniature model of the urban transport infrastructure. But, unlike other elements of transport infrastructure, planning ITTH is not provided by any regulatory documentation.

The present work proposes the review of the main regulatory requirements for ITTH urban design taking in account that every element of transport infrastructure such as public transport, road network, pedestrian commuter etc., needs the same set of regulatory documents to be developed.

[a] Corresponding author: vlasych@mail.ru

The urban planning Code of the Russian Federation [1] provides the development of regional standards for urban design. They must define «a set of indicators for minimum possible provision on the objects of regional importance". Besides, the Code provides the development of similar local standards.

Regional urban planning Codes can include the development of other types of regulatory documents. For example, the Urban Development Code of Moscow [2], in addition to Regional Standards, provides the development of Guidelines for Urban Design in the Moscow City.

Thus, in accordance with the current legislation, there is a growing need in development of several types of regulatory requirements and guidelines for the design of Transport Transit Hubs, such as:

- Regional Planning and Urban Design Standards;
- Guidance on Urban Design;
- Design Manual for Intermodal Transport Transit Hubs.

3 ITTH regional planning and urban design standards

The regional regulations on the development of Transport Transit Hubs System should contain the following sections:

- *Glossary* containing the definition of the key terms used in regulatory documents;
- *ITTH Classification* and
- *Service Quality Indicators*, comprising:

a. Transfer Guarantee Time.

Transfer Guarantee Time indicator allows objectively assess the passenger service quality. Paragraph 6.31 of the Federal "Guidance on Urban Planning..." [3] regulates that the transfer time at ITTH should not exceed 3 minutes.

b. Maximum transfer distance.

Foreign standards [4] tell that the maximum transit distance depends on the type of the change and averages:

• 150–160 m in case of change from land public transport modes to high-speed transport modes;

• 460–470 m in case of change from individual mode of transport to high-speed modes.

The current Russian Guidance [3] regulates that the walking distance from the stops of land public transport (p. 6.29) must average 500 m.

c. The level of passenger service (Level of Service – LOS).

This indicator is not applied in national planning practice. In foreign practice it is an integrated index based on performance measure like intensity, speed, density, etc. of pedestrian traffic. LOS is calculated separately for each main element of the hub. The application of LOS indicator in the Russian practice is impeded because of a lack of representative surveys of national ITTHs.

d. Hierarchical access control.

Under the hierarchical access control it is defined the priority access system to the ITTH territory. In American practice (example the district of Columbia, for example) it is used the following hierarchy (starting from top level): "pedestrians – bicycles – road passenger transport – individual transport arriving for drop-off – private transport arriving at "Park and Ride". Such a hierarchy is associated with trips pattern to metropolitan stations in the Columbia District, where the ratio of road passenger transport arriving on transport transit hubs averages 15%. For the realities of national life seems more appropriated to establish the following hierarchy: "land public transport – pedestrians – bicycles – individual transport for passengers landing – individual transport for all types parking".

One more alternative is the establishment of different hierarchies for the various planning zones of the city. Thus, the issues of the hierarchical access control become key ones to ensure rational and effective ITTH functioning and deserve the most careful study and discussion in the professional community.

4 Instructions on urban design

It seems appropriate to highlight the following main parts of the Guidance.

Introduction, to be devoted to the main issues of ITTH development.

It should be noted that the current practice [5–8] allows distinguishing three fundamentally different approaches to the urban design of transit hubs.

The first approach can be defined as a reconstructive-organizational development method. The essence of this method is the integrated local reconstruction of the ground level station areas only. In Russian planning practice, such solution is called "planar interchange node".

The second approach can be defined as a transport infrastructure priority development method. It suggests the exclusive construction of transport-purpose objects within the territory of the hub, mostly of "Park-and-Ride" facilities. At that, it is necessary to ensure spatial and functional relationship of all main elements of the site, especially between "Park-and-Ride" and landing platforms on high-speed stations [9]. In addition, the implementation of this method suggests also the realization totally or partially of the local reconstruction foreseen by the first method.

The third approach can be defined as a global construction method, as it involves the construction of a multifunctional interchange node on the territory of a transport transit terminal. The main aim of the interchange node is to provide comfortable transfer from one mode of transport to the other. Therefor it is suggested to build a single spatial ensemble that will unite under one roof all the main elements of a transport hub. Besides, in order to attract the investors for funding of the construction and for further development of the transit terminal it is suggested to give accommodation within its structure to residential, entertainment and retail areas intended for commercial leasing.

The implementation of any of these approaches should contribute to the realization of the main social domain of any kind of Transport Transit Hub, such as to ensure the most comfortable transfer conditions while changing transport mode with minimal time costs, including also the possibility for the provision of full range of associated passenger services [10]. The need of the transport transit hub reconstruction is determined by the level of its conformity to the norms and regulations of the Regional Standards and Guidance on Urban Design.

The above mentioned approaches can be realized separately or as the sequence of Transport Transit Hub reconstruction. As an example, let's consider the development of the Moscow ITTH "Vodny stadion".

On the photo (Figure 1) it is presented the current state of things on the ITTH. Even a cursory look at the photo allows to evaluate transfer conditions as not satisfactory, which effects the public service quality indicators, lowers comfort and safety of passengers stay on the ITTH territory, of the TPU, complicates functioning of all modes of transport.

Fig. 1. The current state of the ITTH territory "Vodny stadion"

Figure 2 presents the composition parts of the project structure of the hub and proposals for a coherent evolutionary reconstruction of the ITTH territory.

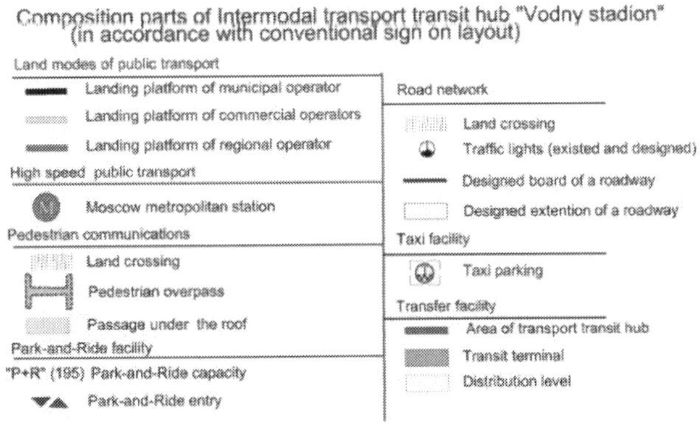

Fig. 2. Structure of ITTH «Vodny Stadion», Moscow, RF

Figure 3 shows the first simplest approach of planar reconstructive and organizational development of the ITTH territory.

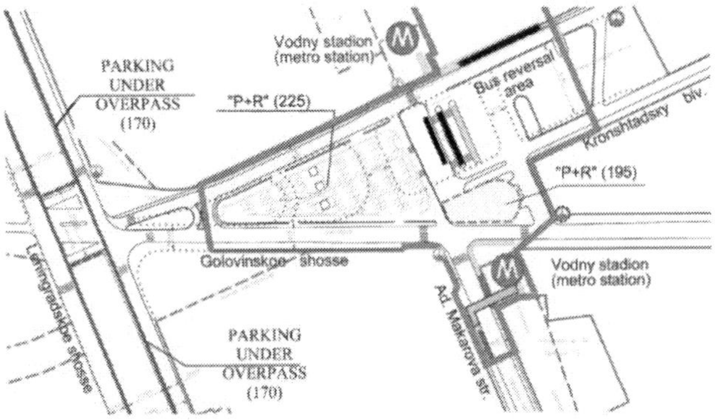

Fig. 3. ITTH «Vodny Stadion». The planar project in the ground level.

In the course of evolutionary development of the site it is proposed the construction of local terminal only, to organize transfer process between land public and high-speed transport modes. It includes landing platform on the ground floor, distribution level on the first floor and pedestrian connection with the metro station (Figure 4).

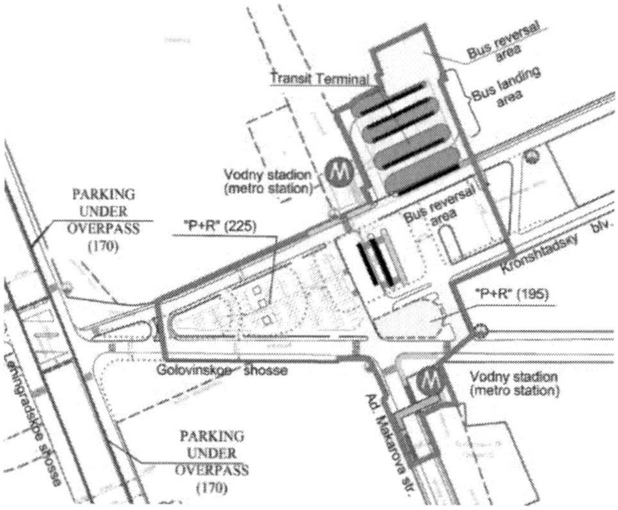

Fig. 4. ITTH «Vodny stadion». The project of compact transit terminal for land modes of public transport.

The large-scale level of the ITTH development includes the construction of huge multifunctional transit terminal with complex transport and public functions (Figure 5).

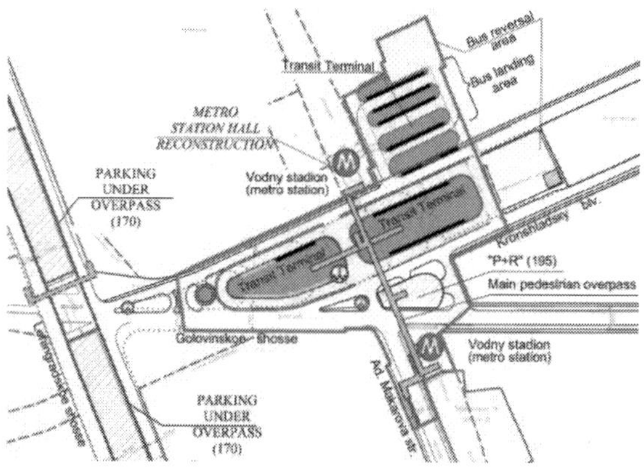

Fig. 5. ITTH «Vodny stadion». The project of huge complex transit terminal (superior development level)

Requirements for TPU. Every design solution for construction or reconstruction of any ITTH should follow three main principles, such as:

– ensuring passenger comfort and safety;

– multifunctionality, which requires compact placement of all ITTH facilities and elements using every possibility to combine different devices and facilities to ensure the best effectiveness;

– Complexity, which requires interconnection not only between all the elements of the site but also between the site and with the adjacent urban territories, with the account of their future sustainable development.

In order to follow these principles, the author suggests to group the basic requirements for ITTH in three categories:

– transport requirements;
– urban development requirements (requirements for composition) and
– technologic requirements (requirements to ensure the efficiency of public transportation systems).

5 ITTH design manual

ITTH design manual should contain standards for calculation of all main areas of the hub, which include:

– communication zones;
– public transport areas and infrastructure;
– city road network and network of ITTH driveways;
– parking facilities.

All design activities must proceed in accordance with the evolution planned phases of the ITTH development process: for the period of the hub start-up; for the first state of development (5–7 years after the reconstruction); and for the estimated service life (20–25 years). For sustainable operation, the ITTH should have 10–15% of reserve capacity of all standard elements [11].

6 Conclusion

The above shows the urgent need to development the regulatory base of ITTH design, reflecting their urban and socio-economic specificities, defining rules for their construction, reconstruction and operation. In this regard, the author considers it appropriate to develop and approbate the following regulatory documents:

– Regional standards for urban design of ITTH;
– Instructions for ITTH urban design;
– ITTH design manual;
– The calculation methodizes of the basic elements of ITTH.

Testing a complex normative documentation is available at transit nodes of Moscow and Moscow region.

Improving the regulatory framework and the functioning of ITTH requires performing complex of scientific-methodological research, which will justify quality indicators of ITTH transport services, allow the rationing of transit time, determine the minimum required composition of ITTH infrastructure elements.

References

1. The town-planning code of the Russian Federation from 29 dec. 2004, 190-FZ, http://base.garant.ru/12138258/#ixzz41gbpboki (2015)
2. The Moscow town-planning code, No. 28 from 06.25.2008 (2015)
3. SP 42.13330.2011 the Federal Guidance on Urban and Rural Planning, http://www.minregion.ru/activities/771/tehreg/482/484/486/1045.html (2015)
4. Station Site and Access Planning Manual, Washington Metropolitan Area Transit Authority, https:// www.wmata.com/pdfs/ planning/ Station%20Access/ SSAPM.pdf (2012)

5. D.N. Vlasov, *Transport transit hubs of the biggest city (on the example of Moscow city)* (2009)
6. D.N. Vlasov, *Scientific and methodological basis of transit transport hubs system development (on the example of Moscow agglomeration)* (2013)
7. D.N. Vlasov, Academy. Architecture and construction, **3**, 86–89 (2013)
8. D.N. Vlasov, Real estate: economy, management, **2**, 55–58 (2011)
9. N.V. Danilina, *Scientific-methodical bases of formation of system of "Park and ride" Parking areas in the largest cities: the example of the city of Moscow*: the dissertation on candidate of technical sciences degree (2012)
10. N.V. Danilina, Internet-Vestnik VolgGASU, **2** (**33**), 29 (2014)
11. E.V. Sherbina, N.V. Danilina, D.N. Vlasov, Jour. of Applied Engineering Research, **10-22**, 43131–43138 (2015)

Adaptation of urban roads to changing of transport demand

Sergey Ertman[1a], *Juliya* Ertman[1], and *Dmitriy* Zakharov[1]

[1]Automobile Transport Operation Department, Transport Institute, Tyumen Industrial University, 625000, Tyumen, Russia

Abstract. To realize the paradigm of the "Sustainable city" it is necessary well in advance to evaluate the ability of the transport system to meet the raising transport demand to decrease the city network load and reduce the influence on the environment. The index of the network element adaptability can exist as such an evaluation. This article contains general results and conclusions of the investigation, had been carried out on the section of the Tyumen city highway to reveal and predict the transport demand to evaluate the adaptability of the highway elements to meet the demand outlook.

1 Introduction

The characteristic feature of nowadays is the increase of life rate. Meanwhile the bigger the megalopolis is, the denser people, transport and information streams are and the bigger the necessity in increasing its velocity is, the more difficulties they experience when moving and the bigger, correspondingly, the time in journey and the possibility of mistakes at their interaction are.

This is natural. That is why the paradigm realization of the "sustainable city" presupposes, first of all, improvement of the methods to control information, transport and pedestrian streams to provide the very possibility of their movement and effective interaction when minimizing their influence on the environment [1–3]. One of the necessary elements of such control is prospecting the load change on the city infrastructure to correct resources planning and prevent the stress situations [4].

2 Methods of research

2.1 Transport demand determination

In the field of transport streams control there is the notion of transport demand, which is defined as meeting the needs of people in movement. It is caused by the space division of different requirements [5, 6]. Correspondingly, the existing transport infrastructure and its possibility is named as transport supply [5]. In the accounting methods of prediction

[a] Corresponding author: ertmans@list.ru

the problem of determining the future movement intensity is known as well as correspondingly the expenditures to overcome the network elements under the known parameters of transport infrastructure.

Transport demand is characterized by the time and space irregularities. The irregularity of transport demand is influenced by the seasonal factors, changing of the objects of absorption and formation of transport and people steams, schemes and methods to organize their movement and etc. different elements of the transport network have different level of carrying capacity under the influence of one and the same factors. Thus, the elements of transport network possess the adaptability properties to the variable factors of the environment [7].

Adaptability in this context is the property of the transport network elements to keep the values of their characteristic indexes (carrying capacity) on the nominal (optimal, acceptable) level when changing the influencing environmental factors (transport demand) from their standard meanings.

The notion of demand, supply and adaptability is topical for any material and people streams and transport networks. We will stop at the task of determining the adaptability of the city road network to the predicted increase of transport demand, irregular by its nature.

2.2 The matrix of transport correspondences

One of the methods to determine the value of the future transport demand is the superposition on the existing or projected transport network of transport correspondences, which are realized by the different ways of journey. Forming the matrix of correspondences is the central and the most complex task, from the mathematical point of view, in all investigations, connected with the essential structural or parametrical changes of transport streams of the city network.

The method of decomposition was used as the method to evaluate the matrix of transport streams correspondences, which is sensitive to the changes of the initial correspondences matrix, of low and top limits being evaluated and artificial variables, which gives the possibility to use it for evaluation of different scenarios of network development and hypotheses of the stream intensity change.

In the method of decomposition the division on the matrixes of correspondences of the smaller order is used so, that the distribution of the correspondences of each of these matrixes is described by the principle "everything or nothing". The given matrixes, used as the evaluation of each of the newly formed correspondences matrixes and describing the belonging of the correspondences to the graph arcs, are binary. This simplifies the preparation of data, their control and visualization. Compensation of the work content increase at increase of the number of the correspondences matrixes being evaluated is the reduce of dimension of each of them.

With the help of the method to restore the matrixes of correspondences it is possible to evaluate the network functioning on the basis of only one kind of the initial data – measurements of the stream intensity. In the described investigations the empiric data were used, which were accumulated as the result of movement intensity research on the units (crossroads) and stages of the city road network during the number of years.

The initial data to evaluate the correspondences matrix were presented in the form of oriented graph with the movement intensity values. The description of the network is used at which the correspondences, realized in different ways, are divided into the set of separate matrixes.

The method set allows to evaluate and predict the influence of different city-construction decisions on the size of transport demand under the conditions of heavy traffic in the city.

3 Empiric research

3.1 The section for research

To analyze the adaptability of the road network elements to the changes of transport demand under the influence of time and city construction decisions factors in the city of Tyumen the section of the highway Staryi Tobolskyi tract was used as an example.

For nearly small period of time on the section of Staryi Tobolskyi tract several decisions of the city construction value were realized, and a number of large objects, attracting transport, cargo and passengers, and pedestrian streams were put into operation (Figure 1):

- "UMMC-Steel";
- JSC Antipinsky Refinery;
- Logistics stocks "VELES"
- Out of town cottage settlements;
- Supermarket "Lenta".

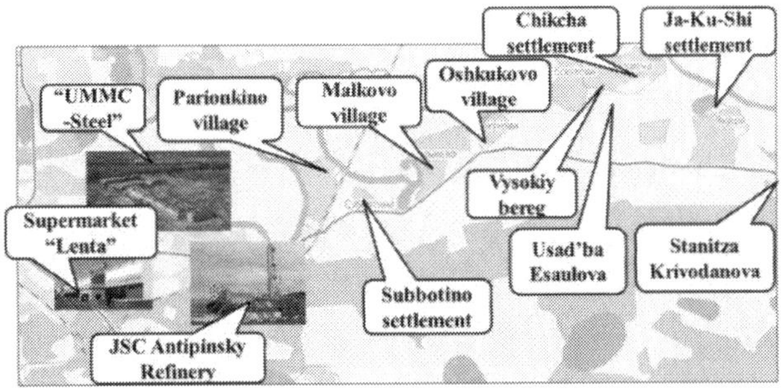

Fig. 1 Summarizing scheme of the section of Staryi Tobolskyi tract.

The section of Staryi Tobolskyi tract at the moment of the investigations beginning (2010 year) had only one possible exit to the city through the intersection of the streets Staryi Tobolskyi, Yalutorovskyi tracts, Respubliki street.

Thus, any change of the range of city and suburban traffic of the streets Staryi Tobolskyi, Yalutorovskyi tract found its reflection in the changes of the traffic intensity at this intersection, it can be with negligible admissions considered that transport demand for the movement along Staryi Tobolskyi tract and out of it was equal to the intensity of forward and backward directions on the pointed out intersection.

Thus, the collection of the transport streams intensity data was being held since 2010 till 2014 just at this very section of the network.

The method of making observations to determine the actual hour intensity of traffic is described, in particular, in the work [8, 9].

3.2 The transport demand exploring

Maximum average-day values of transport demand, reached at the process of observation, are presented as cartograms for 2010 and 14 years on Figure 2.

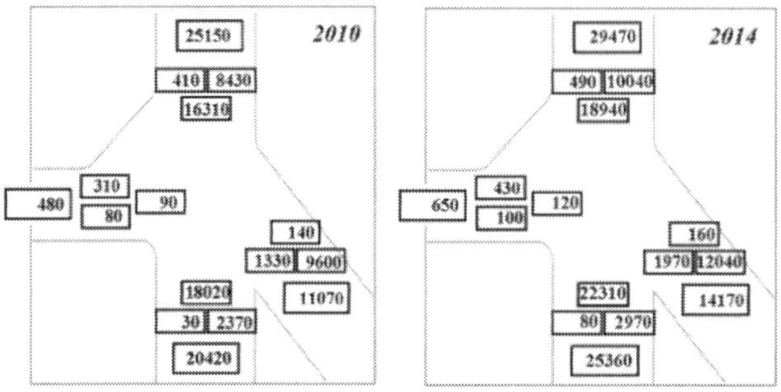

Fig. 2. Cartogram of the maximum average-day values of transport demand on the intersection, reached in 2010, 2012 and 2014 years.

According to the actual values of the traffic intensity and composition, the intersection simulation model was created. It is presented on Figure 3.

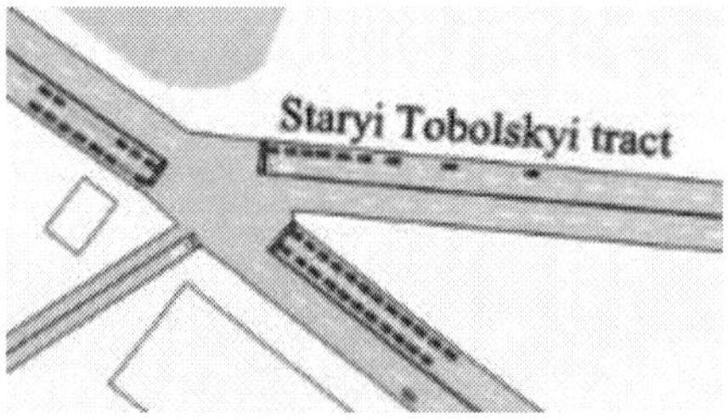

Fig. 3. Intersection simulation model (according to the data of 2012).

3.3 The correspondences matrixes of transport streams

Together with making the investigations of the traffic intensity on Staryi Tobolskyi tract, to determine the characteristics of the population mobility, the main purposes of movement were set as well as main points of absorption and regeneration of transport streams (the so-called base sections).

Actual values of average-day, generating and absorbing abilities of the main points were determined. The values of the transport streams correspondences obtained were put over the transport network, being realized by different ways of destination to (and from) the base sections. Thus, each of them was considered as a separate correspondence.

The data obtained allowed to get a set of binary correspondences matrixes of transport streams along the whole extent of the road section under consideration.

To get the general matrixes of transport correspondences all routes of traffic on the section of the highway Staryi Tobolskyi tract were presented in the view of the orientated graph [5, 6] (Figures 4–8).

Fig. 4. Section of the road Staryi Tobolskyi tract 1–3 km.

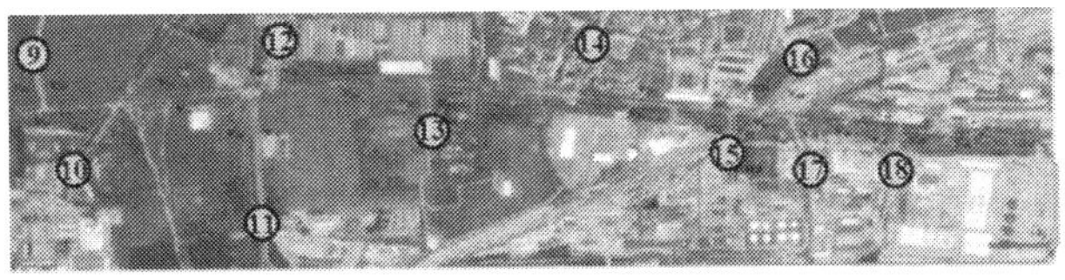

Fig. 5. Section of the road Staryi Tobolskyi tract 3–6 km.

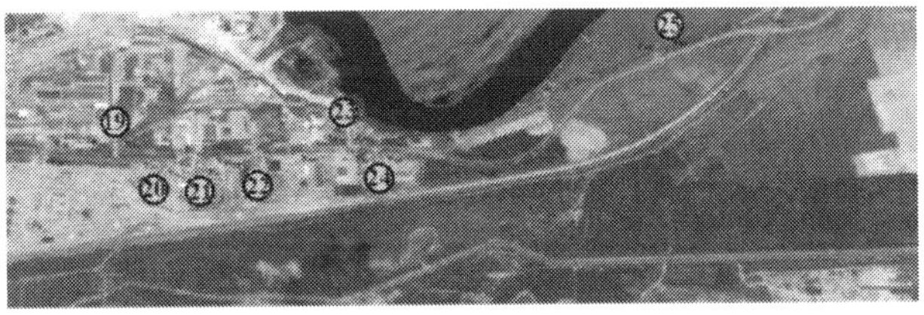

Fig. 6. Section of the road Staryi Tobolskyi tract 6–10 km.

Fig. 7. Section of the road Staryi Tobolskyi tract 10–13 km.

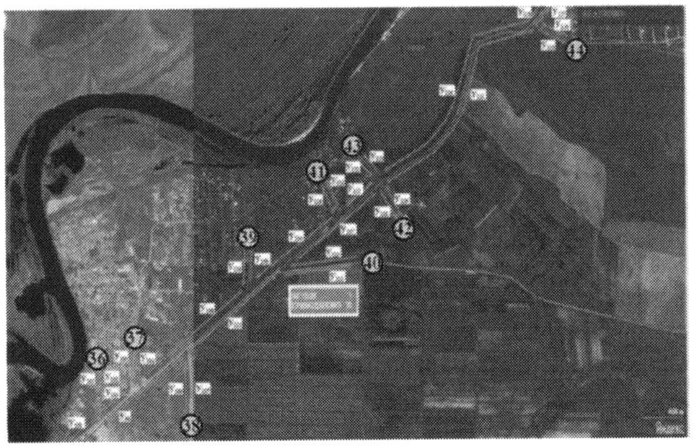

Fig. 8. Section of the road Staryi Tobolskyi tract 13–20 km.

Thus, on Figures 4–8 the section of the highway of Staryi Tobolskyi tract is presented by the oriented graph, including 44 corresponding tops and 128 arcs.

On the basis of the data obtained transport correspondences matrix was determined. Unfortunately, because of a big volume of the correspondences matrix obtained, it cannot be presented in this very article.

3.4 The predicted estimates

For each separate correspondence, connected with a base section, hypotheses of changing the average-day and annual density of traffic were formulated on the basis of the known time trends of occupation the territories, dedicated for the construction (for cottage settlements), productive plans of enterprises development (for “JSC Antipinsky Refinery” and "UMMC-Steel"), growth of birthrate and auto mobilizing the population (for Antipino settlement, Oshkukovo village, Esaulovo settlement, Malkovo village, Subbotino settlement, Chikcha settlement, etc.)

The predicted estimates of increasing the population mobility obtained allowed to determine the hypothetic value of transport demand for the movement along Staryi Tobolskyi tract for the middle and long-term perspective and model the scenarios of road traffic movement condition in different elements of the road network section under consideration. On Figure 9 an example of the transport demand model is presented in 2018 under the 2012 data of the carrying capacity of the key intersection.

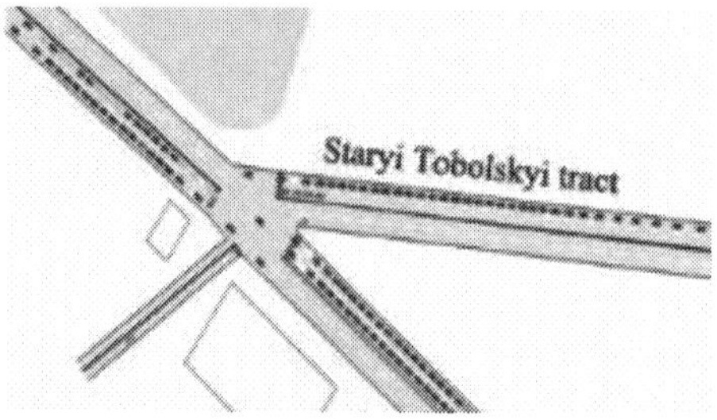

Fig. 9. Model of transport demand at the scenario of development for 2018.

As a result of a comparative analysis of the predictive estimates of increasing transport demand, obtained when using the correspondences matrix, with the characteristics of traffic movement in the dynamics being observed at the intersection under consideration, the values of statistic characteristics were obtained proving the hypotheses got.

3.5 The estimates of adaptability

The estimates of adaptability of different elements of the highway section under consideration were obtained at the scenario of increasing transport demand for 2018 (Figure 10).

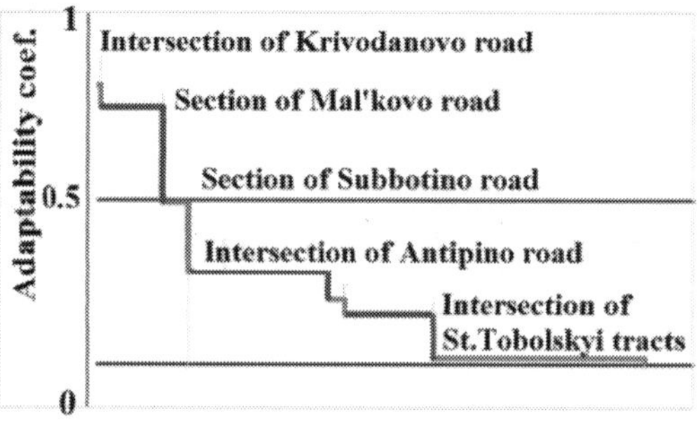

Fig. 10. The estimates of adaptability of different elements of the Staryi Tobolskyi tract road.

At increasing transport demand to the values of 2018 adaptability of the majority of highway section elements is reduced to the values, close to 0. Thus, without the improvement of the road network, realization of the scenario to increase transport demand would be impossible. City construction plans to change the schemes of road traffic organization at the section of Staryi Tobolskyi tract were realized in 2014.

Transport model of the section with taking into account the changes made by the scenario of the development till 2018 allowed to determine that the values of the adaptability coefficients of the intersections increased in comparison with the estimates, obtained without changes, to the average values.

However, implementation in the transport model the scenario of a long-term transport demand development till 2030 again gives reduction of coefficient values of adaptability to critical values.

4 Discussion

Thus, it is necessary to make control decisions beforehand to increase a carrying capacity of the road network section elements on Staryi Tobolskyi tract and their realization to 2025. The ways of raising the carrying capacity at present are the subject of discussion.

5 Conclusions

The results obtained in the course of making investigations, allowed to make the following conclusions.

1. Getting the predicted estimates on changing transport demand and parameters of road traffic under the influence of putting into operation large city construction objects is possible far before the transport collapse occurrence at separate sections and is necessary under the conditions of paradigm realization of the "sustainable city" [2, 4].
2. Determination of the transport correspondences matrix, and the means of simulation modelling allow to get the whole picture of such a change [5, 6, 10], that gives a possibility to plan technical and organization decisions beforehand to level this influence [9].
3. As an estimate of possibilities of road network different elements to meet the transport demand with taking into account its changeable character, and for the perspective, it is possible and reasonable to use the coefficient of adaptability, having the necessary properties of information value, visualization, comparability.

References

1. G. Schliwa, R. Armitage, S. Aziz, J. Evans and J. Rhoades, RTBM, **15**, 50–57 (2015)
2. M. Kiran, EP, **58**, 347–357 (2013)
3. I. Anisimov, A. Ivanov, E. Chikishev, D. Chainikov and L. Reznik, EMP in 21st century, 685–696 (2014)
4. A.S. Santos and S.K. Ribeiro, CSTP, **3** (**4**), 415–420 (2015)
5. D. Lohse, *Transport supply and demand modeling for private and commercial transport – a model-theoretically survey TU* (Dresden, 2004)
6. A. Comi1, P. Delle, F. Filippi and A. Nuzzolo, ET, **51**, 7 (2012)
7. J. Ertman, S. Ertman, I. Anisimov, D. Chainikov and Y. Chikishev, RJPBCS, **6** (**1**), 1761–1770 (2015).
8. Yu.A. Ertman and S.A. Ertman, Science Prospects, **8** (**59**), 48–51 (2014)
9. Yu.A. Ertman and S.A. Ertman, Science and Business: Ways of Development, **8** (**38**), 7–13 (2014)
10. J. Muñuzuri, J.H.R. van Duin and A. Escudero, Procedia – Social and Behavioral Sciences, **2** (**3**), 6165–6176 (2010)

ICSC Problems and Perspectives of high-calcium fly ash from heat power plants in the composition of "green" building materials

Irina Domanskaya[1a], *Valentina* Oleynik[1], *and* Denis Minyazev[1]

[1] Ural Federal University, Institute of materials science and metallurgy, 620002, Yekaterinburg, Russia

Abstract. The peculiarities of the composition formation and hydraulic activity of high calcium fly ash got by burning of Kansk-Achinsk brown coals have been described. Fly ashes selected in dry state can be used as peculiar (specific) binding agents (minerals) on condition of their preliminary handling reducing nonuniform volume change of the ash stone during hydration and also the expansion range of their application in green building. The cavitation technologies, the organization of the fly ash granulation at heat power plants instead of hydraulic ash removal systems and enrichment by various means to extract rare-earth metals are considered to be the most perspective methods of fly ash conditioning.

1 Introduction

Realization of Green Building conception which is an integral part of Sustainable Development makes provision for the application of ecologically and economically effective building (construction) materials which are named "green" ones. In present Russia the national systems of ecological certification and marking of the construction product are based on criterium approach as to the analysis of their life cycle proper [1, 2]. When certifing building materials the influence of the whole complex of a number of processes upon environment is valued (appreciated), namely: from extraction of raw materials to means of their processing during the manufacturing process up to complete demolition, landfill and possible recycling [3]. Under such approach a lot of artificial materials fail to bear criticism; the technologies of their production are too expensive for both the society and nature as a whole. Nowadays the most widely used binding material all over the world is Portland cement, also concretes and mortars on its basis. The concern of the world community as to the perspectiveness of cement production development, the share of global CO_2 emissions being 5% caused the active search of alternative binding materials [4, 5]. Such innovative binding substances are of great practical importance from both ecological point of view and economy because industrial waste are used to produce them instead of the natural raw materials. From our point of view the most perspective base to produce alternative hydraulic (water resistant) binding materials is thermotechnogenic raw material – mineral waste which have suffered different types of high temperature affect to

[a] Corresponding author: i.k.domanskaya@urfu.ru

form liquid phase and kept their amorphous structure partially or completely as a result of their prompt cooling. Granulated blast furnace slags from metallurgical industry and fly ashes from heat power plants refer to this group. The present work is devoted to the problems of utilization and perspective trends of high calcium fly ashes conditioning resulting from burning of Siberian brown coals of Kansk-Achinsk basin.

2 High-calcium fly ash – is a specific type of thermotechnogenetic raw material for production of binders

Fly ashes is not new source of mineral raw for building industry. For many years they are used in ceramics technology, for fractional substitution of Portland cement in production of concretes and mortars, for creation of geopolymeric and other binders and composites on their bases also [6–10]. Despite long-term experience of study and practical use of fly-ashes, there is still urgent and constant need to develop new methods of their application because of significant variation of their compositions and properties [7]. A specific problem present high-calcium fly-ashes (HCFA), that appear while burning brown coals and some kinds of slate coals (C class according to ASTM C618-12). As usually they contain 1–40%, sometimes up to 60% calcium oxide, including that of in free form, not fixed in chemical compounds (CaO free). The CaO free content in HCFA may vary from 0 to 20% and more. The presence of free lime CaO free in fly-ashes, on the one hand, provides them for binding properties and possibility of hardening after hydration. On the other hand, it causes some ecological and technological problems in removal, storage and utilization processes.

2.1 Composition formation and hydraulic activity of high calcium fly ashes

Brown coals are defined as "young" ones and they are the subject of much current interest due to their superior form from peat to coal. The share of coal reserves in Russia are accounted for nearly 40%. As to the composition of brown coals the organo-mineral complexes in the form of Ca and Mg humate prevail in distinction to coals, but the other metals are rare. Therefore, the mineral unburnt (non-fired) part of the brown coals is represented by assemblage of biogenic origin ashes – the so-called "internal"("parent") ash, its quantity amounts to 70% and the "external" one including the impurities of the enclosing rocks [11]. Complex physico-chemical processes of the ash residue formation take place during the burning process of this kind of solid fuel resulting in: organo-mineral complexes decay and interaction of the elements released with both the gas phase components and each other, also melting, crystallization from the melt etc. The experimental results of the composition and properties of the 17 fly ash probes got by brown coals torch burning at Berezovsky heat power plant (Krasnoyarsk region), also the thermodynamic calculations and the probable reactions analyses by using the state diagrams CaO-Al_2O_3-Fe_2O_3 carried out earlier have shown the following [12]. It is established that the supposed CaO isolation from the organic substance (400–600 °C) and its interaction with sulfate component of gas phase is more advantageous thermodynamically. The Gibbs energy for this reaction in oxidative conditions is 280.69 kJ/mol. Thermodynamically less likely the binding of Cao in the aluminates, silicates and carbonates. With increasing temperature, the possibility of sulfation decreases, and the formation of silicates and aluminates of calcium, judging from the values of the Gibbs energy, increases (Table 1). This is confirmed by the results of chemical analysis (Table 2). The most likely phase crystallizing on melt cooling consisting of more than 33–35% CaO is belite, and gelenite consists of less CaO content. However, the low viscosity indicators of the compositions under investigation in the latter case contribute to the conservation of vitrified state at the moment of the fly ash particles cooling down which is confirmed by the X-ray diagram character and is also indicative mainly of puzzolana mechanism of their hardening when mixed with water. CaO free present in flame burned ashes is likely to be the product of the secondary sulfates decomposition.

Table 1. Data of Gibbs energy calculation [12]

Numbers	Compounds	-ΔGT, кДж / mol/K			
		600	800	1000	1200
1.	CaS04	280.69	243.94	208.11	173.62
2.	$CaCO_3$	82.80	52.47	22.97	-5.79
3.	CaO·SiO2	90.17	90.63	90.97	–
4.	2 CaO·SiO2	130.48	132.29	134.29	136.44
5.	CaO·A2O3	27.19	31.40	36.01	41.11
6.	12 CaO·7A2O3	191.42	234.17	276.75	319.99
8.	2 CaO·Fe2O3	61.13	68.47	75.67	82.62

Table 2. Evaluation of chemical composition of fly ashes investigated [12]

Oxides	Oxides content, %, for ashes obtained at the temperature, °C		
	1600–1700	1350–1400	850–880
SiO2	23.68–32.22	17.05–4.16	10.18–21.80
Kvar, %	13.74	22.68	44.05
Al2O 3	7.69–10.13	9.36–13.02	7.35–10.88
Kvar, %	12.34	13.01	16.24
Fe2O 3	9.35–11.17	6.87-12.70	5.33–12.87
Kvar, %	7.31	23.75	32.47
CaO	33.75–38.82	31.33–43.92	31.87–45.30
Kvar, %	5.81	9.65	12.24
MgO	4.52–6.20	4.58–6.97	5.62–3.71
Kvar, %	12.07	13.78	14.74
SO3	2.61–5.94	2.64-12.62	6.72–15.14
Kvar, %	30.92	46.91	32.35

Thus, it can be concluded that the phase composition and the hydraulic activity of high-calcium fly ash depend on their mineral part and the "external": "internal" ratio of ashes, the degree of grinding, temperature, the way of coal burning proper, oxidation-reduction character of gas medium, the duration of the particles stay in the furnace chamber, and the place of sampling in the system of fly-ash trapping [12]. The hardening of the samples on the basis of the binding material without cement is accompanied by expansion which magnitude depends on the CaO free content, SO_3 and the storage conditions. The reason of the expansion proper is nonuniform changing of ash stone volume because of slow hydration of high temperature lime form (the so called "overburning") and also the screening effect of sulphoaluminate phases forming on the ash glass phase surface [13]. By this means brown coal ashes in spite of the hydration active phases available (CaO free, belite, low-based calcium aluminates, glass-phase) providing their ability to harden cannot be used as binding materials(agents) without the additional processing, neutralizing their destructive potential.

2.2 Practicable (possible) ways to neutralize the destructive potential of high calcium fly ash

A great body of information testify that to gain the destructive potential of high calcium fly ash reduction one can use additional grinding, pre-autoclaving, partial hydration in the granulation process with subsequent grinding of the granules obtained. Manufacturing cost of high calcium fly ash activation with the means suggested according to some scientists seems to be higher than the consumption cost of the activated fly ash binder ready for use in the majority of cases. Therefore, the actual utilization volume of the high calcium fly ash in Russia the yield of which amounts to million tones fails to exceed 5% [14]. In our opinion to solve this serious production process the cavitation technology can be used. The experiments have been carried out demonstrating that the cavitation phenomenon resulting in the formation and collapse of cavities after high-rate run through the system of special obstacles (cavitators) is highly effective for hydrodynamic dispersion and slaking of the CAO free excess in brown coal ashes [15]. We have studied the cavitation effect by ultrasound radiation from magnetostriction transducer having taken fly ash as an example of brown coal burning at Berezovsk heat power plant. The research work has been accomplished at the Chair of Binding Materials and Construction Products Manufacture with the scientific workers of the Atomic Stations and Renewable Energy Sources Chair of the UrFU. The content of CaO free in the fly ash is 15%. It has been established that under the optimum parameters of the ultrasound treatment (handling) it is possible not only substantially to reduce the ash stone expansion by 50% when it was cured, but to increase practically by 2 times (twice) the sample strength hardened in normal conditions. As discussed to get maximum positive effect it is necessary to use the activated water-ash suspension proper after ultrasound handling. Its binding potential reduces with time. The results got testify that cavitation effect of ultrasound waves is accompanied with the processes hydration activation of high temperature CAO free, causes the pozzolana reactions intensification when ash glass phase is involved. By this means the experiments carried out confirm the cavitation processes effectiveness to neutralize the destructive potential of the high calcium fly ashes and the necessity to further investigation of both the ultrasound handling of water ash suspensions for the purpose of their optimizing and the chemical reactions mechanisms taking place in these conditions. This allows the initial data to be justified for the development of the commercial installations-neutralizers of the required power.

3 Ecological aspects of high calcium fly ashes utilization

Besides the technological problems (the availability of high temperature CAO free, the variability of the chemical composition) which fail to allow high calcium fly ashes to be used as binding materials on their own, there exist a number of ecological non-solved so far problems complicating the organization of large scale commercial utilization of this kind waste.

3.1 Special features of high calcium ash slags removal and storing at heat power plants

In the majority of cases the heat power plants of the Russian Federation use coal fuel and solid wastes removal is done by using the hydraulic method, namely: fly ashes and slags are mixed with large quantities of water and in the form of a pulp is transported through the sluice discharge pipe -line to the ash dump. In case of the calcium oxide 'increase in ash slag wastes the hydraulic ash removal systems are often broken down because of pipeline "overgrowing" with carbonate accumulation. The dumps to store ash and slag occupy substantial areas of land which are excluded from agricultural alteration of crops and require great costs (expenditure) to keep them in order and are dangerous to the adjacent territories and water reservoirs [16]. Obviously that the question of issue to solve the problem of ash dumps is to eliminate hydraulic ash removal and transform to alternative "dry" systems of ash and slag transportation and storage by using minimum water [9].To improve the ecological situation at heat power plants the science workers of the UrFU together with heat power industry plants offer the system of ash removal with preliminary high calcium ashes granulation. The possibility of the high calcium fly ash dry selection is allowed and its use as specific binding agent (after proper handling). As to fly ash out-of-use it is suggested to granulate it and in the form of pellets with 10–20 mm in diameter to keep temporary on the open ground [17]. It is due to such "gentle" system of the ash removal which will be the future of the heat power plant burning the brown coals of the perspective Kansk-Achinsk basin for Russia.

3.2 Sanitary evaluation of high calcium fly ashes

Sanitary evaluation of the commercial wastes subjected to utilization in the construction materials industry has to include the whole complex of researches, namely: toxicological, sanitary-chemical, radiological. In addition to the product samples by adding the present type of wastes are to be subjected to the laboratory research [16]. It is established that the brown coals radioactivity of the Kansk-Achinsk basin depends on the deposit, the coal seam depth bedding and, practically, is fully defined by 226 Radii which allows the further production control to be limited by defining only this nuklide [18]. According to the data of many scientists studying the content of the natural radionuclide in the brown coals ash dump wastes (A.G. Andreeva, G.V. Kachaev, S.L. Kraft, S.V. Kurkatov, et al.) in the majority of cases it, does not exceed standardized magnitude for the raw materials – 370 becquerels per kilogramme. Therefore, the brown coal fly ash of the dry selection can be used as the raw materials to produce construction materials without limitations on condition of the proper control and the periodicity recommended in every concrete case. As far as the toxicological investigations are concerned one can admit the following. Toxicology of the ashes in question is due to the water soluble alkali compounds and also heavy rare earth metals. Their content and that of the natural radionuklide depend on the deposit and the depth bedding of the coals burnt. Accumulating in ashdump toxic agents threaten the safety of the adjacent territories [16, 18, 19]. At the same time there are means which are successfully tested in the laboratory condition allowing such valuable metals as lantanum, cerium, ytterbium, helium, scandium, vanadium (patents RU 2293134, RU 2138339, etc.) to be extracted. According to the authors, their extraction can justify expenses to organize the commercial recycling of the dry selection fly ash.

4 Summary

From the stand point of view and careful attitude to the environment, the necessity to utilize high calcium ashslag wastes from heat power stations is beyond question, the most perspective being both high calcium fly ashes of the present yield and those accumulated in the ash disposal, occupying thousand hectares of the cultivated lands and containing million tons of finely dispersed mineral thermotechnogenic raw materials. In each concrete case it is necessary to investigate in detail not only chemico-mineral composition of solid wastes defining the perspective trends of utilization, but also their sanitary characteristics including toxicological, sanitary-chemical and radiological investigations. We suggest to consider all possible variants of this raw material conditioning as the necessary constituent when studying the raw material under review that is preliminary handling to fully realize the technological potential taking into account the requirements of both ecology and economics. As to the high calcium fly ash of the present yield there is no doubt the organization the dry selection is to be done and depending upon the composition (CaO free content, rare-earth elements) the investigation the advisability of enriching (valuable components extracting) and\or the activation by means of mechanical, ultrasound or chemical handling (modification). Only the complex approach to both the burning technology of the Kansk-Achinsk brown coals and the problem of high calcium fly ash utilization at heat power plants can provide safety and deserving replacement of the natural raw materials by the technogenic ones including the alternative binding materials production. To start the development of the concrete brown coal deposit seems to consider it not only as power source, but as the object containing complex assemblage of organic and mineral raw materials and consequently to develop the most effective complex technologies of their recycling.

References

1. E.A. Sukhinina, Internet-Vestnik of Volgograd SACU, **1** (**31**) (2014)
2. A.N. Asaul, *Materials of the conference "Effective models and technologies of resource saving, energy saving and environmental management in the construction and housing and communal services"* (2014)
3. V.P. Knyazeva, *Ecological basis of selection of materials in architectural design* (Moscow, 2015)
4. D.N. Huntzingera, T.D. Eatmonb, Journal of Cleaner Production, **17**, 668–675 (2009)
5. M.C.G. Juengera, F. Winnefeldb, J.L. Provisc, J.H. Idekerd, Cement and Concrete Research, **41**, 1232–1243 (2011)
6. C. Heidrich, H.-J. Feuerborn, A. Weir, *Coal Combustion Products: a Global Perspective*, (WOCA) Conference, in Lexington, KY (2013)
7. Z.T. Yao, X.S. Ji, P.K. Sarker, J.H. Tang, L.Q. Ge, M.S. Xia, Y.Q. Xi, Earth-Science Reviews, **141**, 105–121 (2015)
8. Z. Zhang, J. Provis, A. Reid, H Wang, Construction and Building Materials, **56**, 113–127 (2014)
9. A.I. Kalachev, *Proceedings of the V conference "Ashes from TPPs: removal, transport, processing, landfilling"*, 41–46 (2014)
10. L.M. Delitsyn, Y.V. Ryabov, A.S. Vlasov, Energy saving, **2**, (2014)
11. H.A. Iskhakov, M.M. Kolosova, Bulletin of the Kuzbass STU, **4**, 109–111 (2006)
12. I.K. Domanskaya, V.N. Oleinik, Internationale Baustofftagung. Weimar, **1**, 1071–1079 (2006)
13. I.K. Domanskaya, V.N. Oleinik, Internationale Baustofftagung. Weimar, **1**, 0817–0822 (2009)
14. Y.A. Nikiforov, D.N. Neyland, *Mineral wastes: activation and application in low-rise construction* (Krasnoyarsk, 2001)

15. D.N. Neyland, A.Y. Nikiforov, *Hydrodynamic activation and application of ash TPP* (Krasnoyarsk, 2003)
16. S.G. Andreeva, *Hygienic assessment of waste ash produced during combustion of coals of the Kansk-Achinsk basin*: author's abstract on the competition of graduate degree Ph.D. (Kemerovo, 2006)
17. I. Domanskaya, V. Ufimtsev, F. Kapustin, V. Oleynik, WIT Press: Energy Production and Managemente in the 21st Century, **2**, 1129–1136 (2014)
18. Y.E. Yudovich, M.P. Katniss, *Toxic elements-impurities in fossil coals* (2015)
19. L.N. Adeeva, V.F. Borbat, Journal Bulletin of Omsk University, **2**, 141–151 (2006)

E3S Web of Conferences **6**, 02001 (2016) DOI: 10.1051/e3sconf/20160602001
ICSC 2016

Intermodal system for mobility demand in the realities of the Russian Federation: reality and forecast

Nina Danilina[a]

Moscow State University of Civil Engineering, 129337, Moscow, Russia

Abstract. The article talks about development of transport systems in context of sustainable urban development, transport problems, growing mobility demand and the way how to meet the needs of population of the Russian Federation by providing the right transport system adequate for the particular conditions of urban agglomeration being the reality of life in the Russian Federation. The author underlines the important role of passenger transport and makes all-around examination of the Intermodal Transport System, ITTU and Park&Ride facilities as the basis on which the Russian Federation can build sustainable development of its cities.

1 Introduction

Recently, city development strategies have become the priorities of the Russian government. Extensive urbanization and rapid growth of urban territories naturally and permanently tend towards agglomerations and group settlements and generate constant intensification and complication of multiplied transport links, while traditional improvement efforts do not work. One of the most critical issues remains the problem of urban population mobility management, particularly during the peak period travel-times with the highest proportion of work trips. That's why innovative trends and instruments for modern urban transport system planning and engineering and organization are in urgent need to be chosen, investigated and implemented in a way to guarantee cities the sustainable provision of current and future demands in urban transportation.

2 Sustainable development of urban transport system

The main purpose of the urban transport system is balancing between meeting and managing public demands in traffic services using the infrastructure capacity of transport system. By judging results from program goals, it is considered the degree of compliance of transport services with the aims for city sustainable development and mobility of its inhabitants. The Town Planning Code of the Russian Federation determines the general direction of sustainable development of territories as "to ensure the implementation of urban development activities for security and favorable conditions of human life,

[a] Corresponding author: nina_danilina@mail.ru

limiting the negative impact of economic and other activity on the generations" [1]. In the author's view, these conditions can be applied also to sustainable development of urban transport system integrally and to the public transport in particular. Thus, social, ecological and economic directions are the domains for urban transport development in the Russian Federation, which complies to three-dimension model of Sustainable Development Concept [2], where each of three aspects defines the consummation of basic – transportation –function of the system.

The economic domain assumes that solutions in the sphere of transport services to increase urban mobility, accessibility, internal and external transport links are based on knowledge and objective regularities and will bring maximum economic efficiency in the development of the city and, as a consequence, the well-being of the citizens.

Social domain establishes the ensuring quality of life in the aspect satisfaction with population needs in traveling for various purposes. Public transport system offers a wide range of public services and their quality directly affects on the amount of their consumption and the mobility of the population in whole. Mobility right is one of the human and constitutional rights that encompasses the right of individuals to have an access within the territory of the city and the country. The essential conditions are not only the availability of public transport services, as well as high travel requirements to the security, comfort, warranty and planning possibility [3].

The ecological domain stresses the need to ensure environmental safety of the city ecosystem not only here and now, but also with the prospect for the future generations. Motor vehicles are a major source of harmful emissions into the atmosphere, noise, dust, vibration, which adversely affects both natural environment and health and psychological state of the person. The term "sustainability" in relation to the transport development encourages development of engineering, IT, design, spatial, technical and technological activities, including those at the stage of transport planning, up to the international environmental standards aimed at preserving the environment.

Synthesis of aspects defines vector of development of transport services and justifies for the adoption of a sustainable design solution

3 Objectives for the sustainable development of public urban transport system

World experience of sustainable transport systems planning and design marks the need to ensure unimpeded access to any public urban area for all social groups of population and, moreover, to provide the most easy, convenient and safe trip regardless of its purposes [4]. Traditionally, difficulty accessing urban area is measured as time spent on the trip, which can be used as the main quantitative indicator of public transport system development level. This approach permits to judge the sustainability level of public transportation by its ability to offer minimal difficulty trips conditions, regardless of the placement of residential areas within urban areas. As a rule, the urban population grows at a faster rate than the expanding of road network and development of public transport systems. This causes the imbalance between the level of public demand for transport services and potential capabilities of the transport systems. Therefore, the use of the passenger public transport and in particular high speed types of public transport is considered to be the most rational and promising way to trip as it does not depend on traffic conditions on the road network.

The policy of priority development of high-speed modes is fully consistent with the concept of sustainable development [5]. It allows meet the demands of the population mobility with maximum efficiency in land use and transport resources and much less pollution than road transport. The world practice witnesses that formation of comfortable, fast and safe system resolves the problem of public transportation and also generates further urban development in areas of accessibility to its route network stations. There are clearly expressed relations between residential density, travel patterns and distance to public-transport facilities, for example to the main station of high-speed public transport system (Figure 1). Closer is the station, greater is number of the buildings, residential density and number of people eager to prefer passenger transport to trip. Farther is the station, more are the trips by means of private vehicles, due to the increasing difficulty of passenger transport trip. The problem of equitable access of all social groups of population to urban areas and transport infrastructure is most acute in major cities with developed suburbs and in agglomerations characterized by peculiar spatial scale resettlement forms. So the choice of such form of public transport system that balances the difficulty and functional efficiency of the trip regardless of the distance to the public transit stations becomes particularly important.

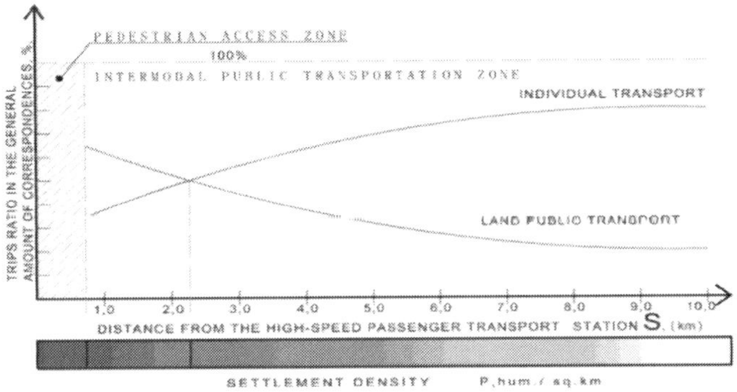

Fig. 1. Relationships between urban density and travel patterns.

4 Choosing right public transport system for the Russian federation

Urban agglomeration is the reality of life in the Russian Federation. Typically, structure and forms of organization of transport system in Russian cities are defined by the number of people living on their territory. Threshold population of the city over 500,000 people suggests priority given to off-street modes to be implemented as the framework of public transport. The cities with population less than 1 million prefer LRT systems or bus rapid transit (BRT systems); in cities with population of more than 1 million it is considered the necessity of construction conventional underground metropolitan or S-Bahn with greater carrying capacity. High-speed modes of transport as part of the urban transportation system become concentration points of passenger load, since, as a rule, they connect main residential areas of the city and its focuses of labor gravity. Due to natural self-regulating urban development processes, vital activity of population centralizes in areas of accessibility to high-speed modes of transport resulting in the development of residential and public functions of adjacent territories.

Current trends in the public transportation enable not only quantitative and qualitative development of actual passenger transport, but of new modes of passenger transportation. According to the author, the most suitable to Russian realities way to meet the mobility demand is the adoption of intermodal communication supported by the system of Intermodal Transport Transit Hubs (ITTH) that already has experienced implementation in the Russian Federation [6]. For example, Moscow Industry Development Scheme of ITTH up to the year 2020 envisages the implementation of 255 ITTH on the territory of the Moscow agglomeration based on subway and railway systems [7]. Official web site of the Committee for transport infrastructure development informs that it has been already decided to construct 11 ITTH in St. Petersburg and that the best pioneer practices of the major cities will be implemented in other developing cities in Russia. In view of such promising development prospects, it seems possible to propose the original Model that forecasts the potential of ITTH and suggests its most efficient and effective use [8].

Research and analysis of ITTH global experience in urban conditions similar to those of the Russian Federation, allows to suggest the development of TPU as urban and suburban focal points with maximum passengers concentration, places of attraction for public and private road transport. Access control to ITTH becomes a demand for managing centripetal process of traffic and pedestrian flows [9]. The main objective is to satisfy mobility demand providing opportunities for free use of transport services offered by the ITTH for people living around and within its zone of influence. It is possible to identify three main areas distinguished by the distance to the station and access mode to ITTH (Figure 1, 2), such as:

- R = 700 m – pedestrian access area;
- R = 2200 m – public transport access area with high proportion of passenger land transport modes users and
- R > 2200 m – individual transport access area with constantly increasing proportion of private vehicles used in direct ratio to the distance from ITTH (authors' study indicates radius about 10–15 km.)

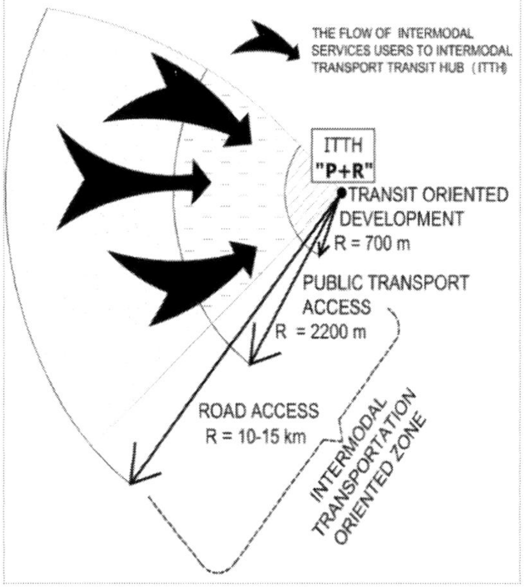

Figure 2. Access conditions to Intermodal Transport Transit Hub.

Besides the distance to the station, each zone is characterized by speed and time of travel time to ITTH both depending on type of the transport used. Under the highest level of development of the system of transport services shall be understood the state of things when a trip between a point I located within n-th influenced zone and a point J located in ITTH by any mode of transport including walking, driving and public transit, is made by the shortest route S_n with the highest speed V_n possible for the L mode of transport, and the less overhead costs T_n, for L mode of transport. If the time T for the part of the trip made after taking high-speed mode of transport assume as a constant value for all groups of passengers, then the very small amount of time to commit travel is determined by mathematical expression:

$$\overline{T}_{IJ}^{L} = \min\begin{Bmatrix} T + S_1^L / V_1 + T_1^L \\ T + S_2^L / V_2 + T_2^L \\ T + S_3^L / V_3 + T_3^L \end{Bmatrix}, \text{[minutes]} \tag{1}$$

The value of the maximum level of development of the system will be influenced by technical characteristics of vehicles and the expected characteristics of the trips, such as speed and overhead time defined by existing or planned urban transport conditions.

Let's consider the specific features and peculiar requirement for transport service of each zone.

The zone of pedestrian accessibility to ITTH bounded by the radius R = 700 m is the area of Transit oriented development policy implementation [10]. TOD policy is internationally recognized as the best way of public transport organization in compliance with the principles of sustainable development in all respects. Walking distance from the area of high-speed passenger transport facilities and stations means minimum need and, as a consequence, minimal time and money trip costs. Transport independence solves a complex of everyday problems and directly affects the quality of life and ecological purity of environment. Economically, territories adjacent to ITTH have strong investment appeal. Transit oriented development areas within ITTH walking distance provide for a diverse and intensive mix of residential, commercial, public, administrative and cultural land users and are characterized usually by high density and high rise building. Sustainable public transport system in such zones requires a dense pedestrian communication network connecting ITTN with places of residence or places of employment in the shortest way and functioning also as a public space. In the absence of urban resources to construct a system of high-speed passenger transport providing the density necessary to cover the whole territory of the city by walking zones in pedestrian access to ITTH, there are alternative ways of organizing transport services, focused on the use of one or another type of transport, in particular, to reach the ITTH.

From a distance of more than 700 meters, pedestrian flow begins to wane in proportion to the difficulty accessing ITTH. When human speed is 4.5 km/h, time spent on foot movement begins to exceed the psychological level of 10 minutes, that stimulates the choice in favor of motorized transport modes, so preferences of ITTH users shift towards intermodal trips. Intermodal trip is a combination of several types (modes) of transport, a transfer from one to the other is made in ITTH. There are two main ways of committing the intermodal trip characterized by two basic types of transfer to a high-speed type of transport from public transport modes or from a private car. Researches confirm that the flow capacity of each type users depends on the distance from ITTH. Let's consider for example two main zones with most characteristic features:

a. Zone situated at distance R = 2200 m., requires about 20–25 minutes trip to ITTH by passenger transport, taking into account overheads, at an average speed of public transport about 30 km/h. These rates do not exceed the psychological threshold and allow increasing the number of users arriving to ITTH by terrestrial mods of passenger transport while ensuring unimpeded access to terrestrial passenger transport services and meeting the following requirements:

– requirements for rolling stock: carrying capacity and quality of the rolling stock shall comply with the level of demand from the population and provide a comfortable and safe travelling conditions;

– traffic requirements: traffic frequency, length of routes, speed must ensure the least possible overhead time waiting. A prerequisite is a clear line route schedule.

– requirements for the highway-road network: priority conditions for passenger transport in ITTH; local events in the area of stopping points and on the route;

– requirement for ITTU: arrangement of boarding-landing points, minimizing overhead costs time to enter the TPU and change process itself;

– travel information requirements: the existence of placards with routes, waiting time and related information, including free access to online services.

b. Zone situated at distance R > 2200 m, is the territory where a personal car becomes the predominant mode of transportation due to the high difficulty of the trip to ITTH. As the psychological threshold of time expenses for intermodal trip must not exceed 20–25 minutes, so this area is restricted by a maximum radius of 10–15 km, taking into account the time overhead costs, at an average speed of road traffic km/h. In this case, the promotion of intermodal travel of "private car-passenger transport» type requires along with controlled-access street network and highway access managing – especially during peak hours – the accommodation of Park@Ride facilities [11].

World experience of Park&Ride facilities attests to the relevance of this service to that part of the population, that prefers private car to passenger transport. During the morning peak hours, when the street network is overwhelmed, the need for planning and high difficulty private car trip shifts the priority in favor of speed passenger transport, especially when the direction of its work duplicates the driver's route. Park&Ride services located in ITTU is the link between street network and passenger transport which allows the owner to leave the car for a long time in safe environment. Integration of Park&Ride in ITTU composition – unified system of pedestrian communications, well organized transit, common payment and cheap rate system – are powerful incentives to use the service and, as a consequence, to increase the number of intermodal trips.

5 Conclusions

Knowledge and experience accumulated allow the author to conclude that Intermodal Transport System – of the number of innovation systems proven by international and Russian practice – fits up the Russian conditions of global urbanization and constant growth of private cars. ITTU and Park&Ride facilities create basis for sustainable development of Russian urban territories in spite of the fact that the intermodal trip time today remains exceeds the normative values. The implementation of Intermodal Transport System in Russia will meet the growing economic and social mobility needs, will provide equal mobility opportunities for all social groups, increase the attractiveness of passenger transport modes and reduce human impacts on the environment. That is, to improve the quality of urban environment and the quality of life of the population.

References

1. The town-planning code of the Russian Federation (by the UDC) of 29 December 2004 No. 190-FZ, http://base.garant.ru/12138258/#ixzz41gbpboki
2. E.V. Sherbina, N.V. Danilina, D.N. Vlasov, Jour. of Applied Engineering Research, **10-22**, 43131–43138 (2015)
3. E.V. Sherbina, E.V. Gorbenkova, Scientific Review, **9**, 128–131 (2013)

4. E.V Sherbina, M. Afonina, *Experience in application of antierosion geo-compositional systems for sustainable development of recreational and sports facilities*, 159–164 (2014)
5. Sustainable planning and design of urban transportation system: national report of OON Habitat Programme, http://www.unhabitat.ru/ru/publications/reports/reports (2013)
6. E.V. Sherbina, D.N. Vlasov, Architecture and construction of Russia, **6**, 2 (2013)
7. Urban planning project of public transit transport hubs formation, http:/www.genplan.ru/
8. D.N. Vlasov, Academy. Architecture and construction, **3**, 86–89 (2013)
9. D.N. Vlasov, *Reducing the burden on the road network of Central planning zone from intercity transit of motor transport (on the example of Moscow)*: the dissertation of candidate of technical Sciences, (1999)
10. D.N. Vlasov, Architecture and construction of Russia, **8 (212)**, 20–29 (2015)
11. N.V. Danilina, *Scientific-methodical bases of formation of system of "Park and ride" Parking areas in the largest cities: the example of the city of Moscow*: the dissertation ... of candidate of technical Sciences (2012)

E3S Web of Conferences **6**, 02002 (2016) DOI: 10.1051/e3sconf/20160602002
ICSC 2016

Measurement strategy to study radon source, entry and dilution rates in energy-efficient buildings in Russia

Aleksey Vasilyev[1a], *Ilia* Yarmoshenko[1], and *Michael* Zhukovsky[1]

[1]Institute of Industrial Ecology UB RAS, 620990, Yekaterinburg, Russia

Abstract. Main physical quantities, which control indoor radon in modern energy-efficient buildings of Russia was studied basing on the results of radon survey in Ekaterinburg, Russia. It was approved that measures to increase energy efficiency lead to reduction in the ventilation rate and high accumulation of radon concentrations indoors. Long-term measurements with application of proposed mathematical methods give an improved understanding of the radon entry processes. Combined use of integrative and time series (differential) measurements provides effective way to develop standard protocol for radon measurements in energy-efficient multi-storey buildings.

1 Introduction

In this study, attention will be confined to the problem of population exposure to radon in context of energy efficient construction technologies. With the introduction of high-performance insulating materials and other architectural and construction concepts air permeability of building envelope becomes sharply reduced. Besides effective reduction of air permeability of building envelope, conditions for increase of many indoor-generated air pollutants are created [1, 2]. Radon is the one of the most important indoor pollutant from the standpoint of radiation safety. Therefore, the need to consider the protection of public health from radon exposure in conditions of energy-efficient construction is a major task in the field of radiation hygiene.

Multi-storey block buildings represent the main type of dwellings in modern cities of Russia. There are two important sources that contribute to indoor radon: exhalation from the building materials and from the soil (Figure 1). Buildings materials are supposed to be dominant radon source, as proved by continuous measurements of radon concentration in multi-storey dwellings with typical natural ventilation system [3]. Air exchange rate (AER) between outdoor and indoor atmosphere has significant effect on the radon concentration, but implementation of energy-efficient measures result in an increase of air permeability of building envelope. Conditions for high indoor radon concentrations are created.

[a] Corresponding author: vav@ecko.uran.ru

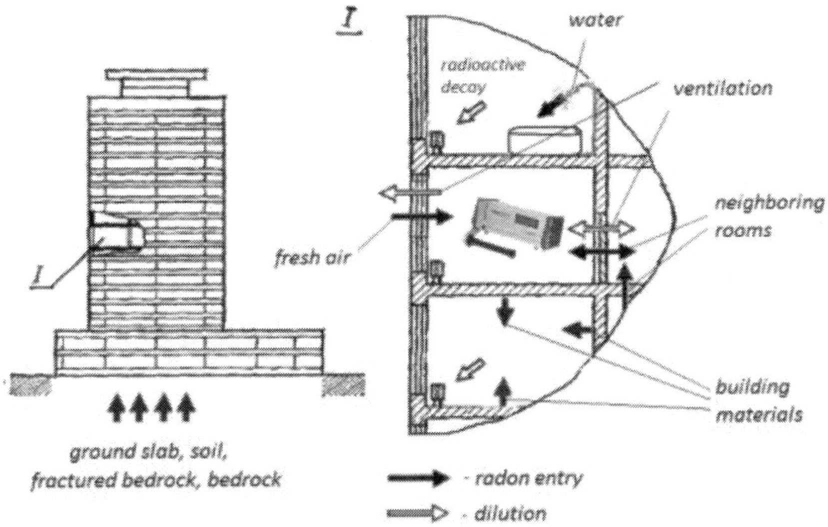

Fig. 1. Indoor radon in multi-storey building: parameters of interest.

The main task of the study is estimation of the main physical quantities, which control indoor radon in modern energy-efficient buildings of Russia, such as properties of radon source, entry and dilution rates.

2 Materials and methods

The sample of dwellings for analyses includes 11 multi-storey buildings having been constructed after 2000 with implementation of the features of energy-efficient construction in Ekaterinburg, Russia. The building characteristics are the following: monolithic reinforced concrete structures, high-performance windows and exterior heat insulating walls of aerated concrete.

A method, that is based on high-resolution radon time series and allows to determine the ventilation rate corresponding inactive and active modes of room use of radon entry analysis, was developed earlier [4]. The method consists of the continuous measurements of radon concentration, temperature and pressure difference between indoor and outdoor atmosphere and analysis. Dwellings were equipped with AlphaGUARD monitors. Measurement interval was set to 60 minutes. Duration of measurements was 6 months. The beginning of the measurements was assigned on winter or summer, so duration of measurements covered cold and warm seasons.

A typical form of radon concentration time series is shown in Figure 2 (upper part). Points are results of 1-hour measurements of indoor radon concentration, estimated AER is shown on the bottom part. It is well known that radon concentration in a room follows a periodical pattern, which can be explained both by meteorological parameters and by living habits (e.g. the tendency to open windows for comfortable conditions). Temporal variations in radon concentration give an opportunity to investigate the radon entry parameters and assess the dominant factors influencing these variations in more details. We suggest to use the autocorrelation function for describing the structure of the temporal dependence from time profile of radon, while the dominant periodicities in the data are examined by means of spectral analysis. Such analysis was conducted applying long-term radon time series obtained using AlphaGUARD radon monitor.

 DOI: 10.1051/e3sconf/20160602002

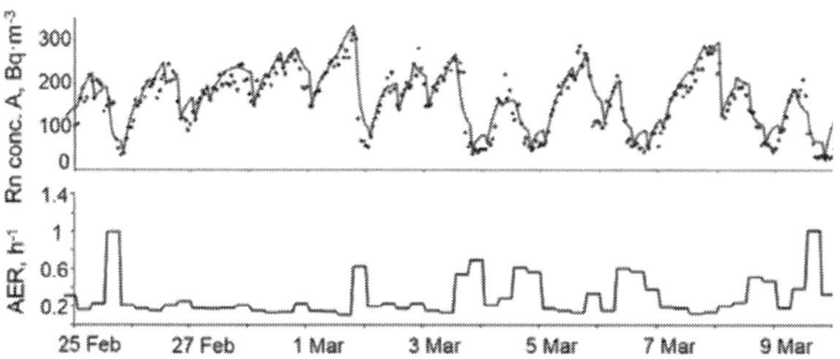

Fig. 2. Time series of 1-hour indoor radon concentration in energy-efficient building and results of preliminary analysis (on the upper part: points are results of 1-hour measurements of indoor radon concentration, curve is model; on the bottom part: estimated ventilation rate).

Software to highlight the sub-series of experimental data characterizing transition from inactive to active mode of room use was developed. Software provides automatic search of the time intervals of decreasing or increasing of indoor radon concentration. Next, the radon entry rate was obtained by the use of regression analysis [5]. Dependence of radon entry rate in a room on indoor/outdoor temperature difference allows to determine dominant mechanism of radon entry – diffusion (absence of dependence) or advective (increase of radon entry rate). Russia is country with a continental climate, typically characterized by long winter and short summer. Indoor/outdoor temperature difference attains value of 50 degrees. Finally, average contribution of the diffusion mechanism of radon entry was defined as ratio of the radon entry rate at zero degree difference between indoor and outdoor atmosphere to radon entry rate at mean temperature difference [5].

3 Results

Results of analysis for typical modern multi-storey buildings in Yekaterinburg, Russia are shown in the Table 1. As shown in the Table diffusion mechanism of radon entry is always dominant. Importance of the diffusion entry in modern buildings is connected with reduced air permeability and lower air exchange rate. According to the results, radon concentrations in modern multi-storey buildings in Russia are relatively high; however, the radon entry rates remain low. Therefore, the main factor leading to higher radon concentrations indoors is low ventilation rate.

Figure 3 presents Fourier spectra (raw periodograms), showing spectral density of the radon concentration as a function of period. The spectra indicates time periods that regulate the radon variations in studied dwelling the following: 12 hour, 1 day, 7 days. Autocorrelation function reveals importance of living habits in radon accumulation (Figure 4).

Table 1. Characteristic parameters of the radon entry and dilution

ID	Floor/number of floors	Mean radon concentration, Bq/m^3	Mean air exchange rate, h-1	Mean radon entry rate, Bq/m^{3}h-1	Ra-226 effective concentration, Bq/kg	Contribution of the diffusion mechanism
1	7/16	203	0.26±0.03	38±2	32	0.75
2	4/16	193	0.32±0.06	42±3	35	0.64
3	3/24	173	0.52±0.05	63±4	53	0.59

4	6/10	143	0.40±0.05	33±1	28	0.92
5	3/16	120	0.35±0.03	38±2	32	0.86
6	12/16	149	0.27±0.09	31±9	35	0.59
7	1/16	133	0.31±0.08	31±7	37	0.67
8	15/25	368	0.27±0.06	65±10	54	0.94
9	17/25	123	0.48±0.08	42±5	33	0.87
10	6/16	99	0.41±0.08	38±2	30	0.95
11	4/16	87	0.35±0.03	23±2	28	0.70

As shown in the Figure 5 cumulative distribution of AER depends on the mode of room use.

It is Important to note that even the floor where room is located doesn't matter (if it is above second floor). As you can see in the Figure 6 there is no dependence on the floor for the building in which measurements were conducted on the different floors (ID 8, Table 1).

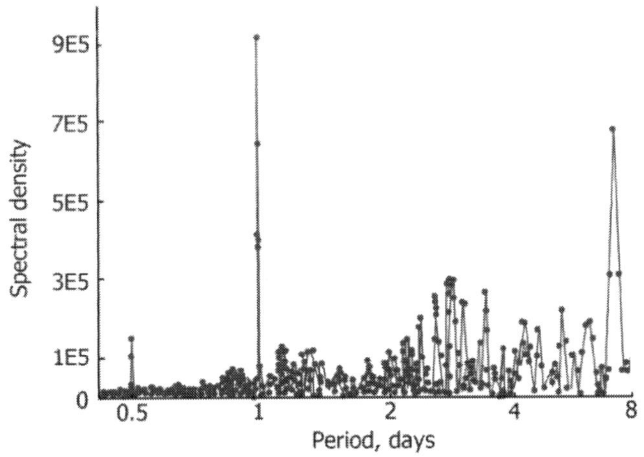

Fig. 3. Fourier spectra obtained from radon profile.

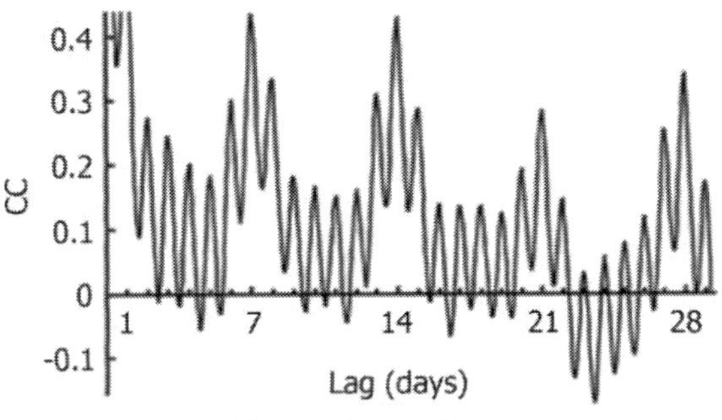

Fig. 4. Autocorrelation function obtained from radon profile.

 DOI: 10.1051/e3sconf/20160602002

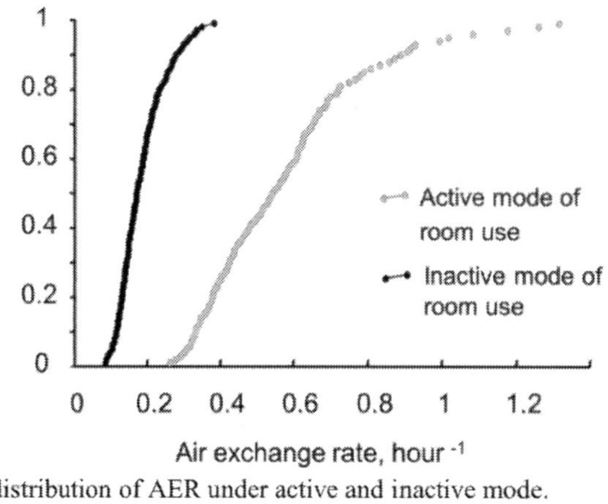

Fig. 5. Cumulative distribution of AER under active and inactive mode.

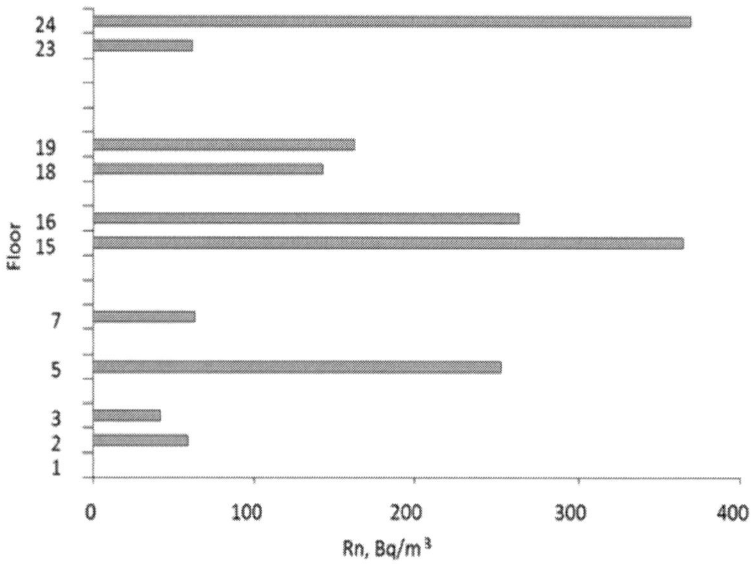

Fig. 6. Variation of radon concentration in multi-storey building (ID 8, Table 1).

4 Conclusion

Based on estimates of average ventilation rate under real conditions of room use in modern multi-storey dwellings, it was approved that measures to increase energy efficiency lead to reduction in the ventilation rate and high accumulation of radon concentrations indoors. Long-term time series analysis with application of mathematical methods (regression analysis, autocorrelation and spectral analysis) provides reliable and efficient technique for revealing radon entry, accumulation and dilution in energy-efficient buildings. Nevertheless, for certain purposes it is not enough yet. For example, the radon concentrations on different floors in multi-storey building are unpredictable even if we know the radon entry and dilution rates (Figure 6). It looks like one still needs to use measurements with short-term passive detectors on the various floors in multi-storey buildings. Despite the existing protocols being reasonable for single-family houses, short-term measurements in multi-storey buildings are still an open question (number of measurements, number of floors to study and etc.) Long-term measurements with application of proposed mathematical methods give an improved understanding of the radon entry processes. Combined use of integrative and time series (differential) measurements provides effective way to develop standard protocol for radon measurements in energy-efficient multi-storey buildings.

References

1. J. Milner, C. Shrubsole, P. Das et al., BMJ, **348**, 1–12 (2014)
2. H.R.R. Santos and M.S. Leal, Energy Build, **54**, 111–121 (2012)
3. I.V. Yarmoshenko, A.V. Vasilyev, A.D. Onishchenko et al., Radiat Prot Dosim, **160**, 53–55 (2014)
4. A. Vasilyev and M. Zhukovsky, J Environ Radioact, **124**, 185–190 (2013)
5. A.V. Vasilyev, I.V. Yarmoshenko and M.V. Zhukovsky, Radiat Prot Dosim, **164**, 601–605 (2015)

Improving vehicle adaptability to the operating conditions of «smart» cities in the northern regions

Ilya Anisimov[1a], *Elena* Magaril[2], *Romen* Magaril[3], *Evgeniy* Chikishev[1], *Denis* Chainikov[1], *Aleksandr* Gavaev[1], *Sergei* Ertman[1], *Yuliya* Ertman[1], and *Andrei* Ivanov[4]

[1]Department of Motor Transport, Tyumen Industrial University, 625000, Tyumen, Russia
[2]Research and Educational Centre ENGEC, Ural Federal University, 620002, Yekaterinburg, Russia
[3]Department of Oil and Gas Processing, Industrial University of Tyumen, 625000, Tyumen, Russia
[4]Technical systems in the agriculture, Northern Trans-Ural State Agricultural University, 625003, Tyumen, Russia

Abstract. The paper reviews a relevant problem of providing a higher quality of people's living in the northern regions by the application of the modern technologies within transport in order to improve its environmental performance and efficiency. The aim of the research is to substantiate methods of increasing the adaptability of vehicles to variable low-temperature operating conditions based on a neural control system of preheating and maintaining the optimum temperature of technological fluids by SHF radiation and improving diesel fuel properties by introducing a multifunctional additive. It has been found that SHF radiation influences the physical and chemical parameters of diesel fuel, as well as the engine output parameters. SHF radiation results in a reduction of fuel consumption to 7%, smoke from the exhaust to 15%, and nitrogen oxide emission to 30%. A conclusion is made that using SHF radiation for technological fuel preheating is essential. An approach has been developed that improves low-temperature properties of diesel fuels and increases engine operation efficiency under severe climatic conditions by introducing a multifunctional additive. The additive will enable a significant decrease in fuel consumption under low-temperature operating conditions and a complex improvement of fuel properties.

1 Introduction

Nowadays, rapid depletion of energy resources and rise in their cost, as well as increased environmental problems necessitates building the urban environment not only in terms of the presence of infrastructure, but also considering its management competences. It is possible to ensure effective control by using modern innovative technologies based on the «Smart city» system, which provide modern life quality. Such technologies shall provide for an economical and environmentally friendly use of municipal life support systems.

[a] Corresponding author: tkcc@list.ru

In a «smart» city, transport based on the intelligent transport system plays a special role. GOST R ISO 14813-1-2011 «Intelligent transport systems (ITS). Reference model architecture(s) for the ITS sector», Part 1 «ITS service domains, service groups and services» provides requirements for vehicle design. Their design must ensure the improvement of safety, reliability and vehicle operation efficiency by cautioning the users or managing the systems or units of vehicles. It should be noted that the quality of fuels ultimately influences the environmental safety and operating efficiency of vehicles. It would be impossible to achieve them with low-quality fuels, even if the design solutions were optimal.

Vehicle efficiency means the ability to perform the required work under given conditions while minimizing the energy costs and exhaust gas emissions.

Thus, «smart» city vehicles must manage their systems and aggregates in order to reduce fuel consumption and exhaust gas emissions.

As far as Russia is concerned, the proportion of urban population is about 75%; the country is at 4th position in the world by the number of cities with a million-plus population (15 cities), the first three being China, India and Brazil. At the same time Russia is the coldest country in the world. Therefore, designing vehicles for a «smart» city should take into account the above-mentioned features, namely a considerable time of operation at low air temperatures that significantly change fuel consumption and exhaust gases emissions [1]. In addition, vehicles react differently to the changes of environment, i.e., they are differently adapted to variable operating conditions. It is necessary to develop approaches and devices to improve the adaptability of vehicles to low-temperature operating conditions [2, 3]. This problem can be solved by applying operational and organizational measures, developing additional units and aggregates that improve the adaptability level and using high-quality motor fuels adapted to low-temperature conditions.

2 Methods of vehicle adaptation to low-temperature conditions

The main directions of research to improve the efficiency and adaptability of vehicles to low-temperature conditions are related mainly to the improvement of structures, technological parameters of transport, technological process organization, as well as with fuel quality modification. The main directions of current research in energy efficiency for the transportation facilities are: optimization of the structure and improvement of the characteristics of the vehicle fleet (including the design of engines and vehicles themselves), improvement of the maintenance system, road network and traffic management, reduction of the negative impact of climatic conditions, improvement of low-temperature fuel characteristics, both by oil refining methods and application of the fuel additives.

Increasing the adaptability of vehicles to variable low-temperature operating conditions can be an effective method of increasing their efficiency. A considerable amount of research [4–7] is devoted to the issues of improving the adaptability, which attract the interest of different scientific groups.

Nowadays, effective methods of vehicle adaptation to low ambient temperature conditions are not so commonly applied in Russia. The known methods of technological fluids (TF) preheating (preheaters, tape preheaters, thermoelectric heaters, etc.) are used and studied. However, more advanced methods of preheating, in particular heating by super-high frequency (SHF) currents, which is one of the perspective methods of vehicles adaptation to low-temperature operating conditions, are practically not applied. The authors witness an increased interest in studying the capabilities of hydrocarbon preheating by SHF currents in vehicles and technological machines (pipeline, railroad transport); however, there is no complex method of applying heat to various TF in a vehicle by means of SHF radiation.

2.1 Using SHF radiation for technological fluids preheating

A reserve of improving the transport operation efficiency is creating a neural network control system of preheating and maintaining the optimum temperature of TF. In the recent past, it was impossible to solve this problem, since there was no componential base of devices capable of instantly getting into gear and providing a uniform heating of the full volume of TF. Application of SHF radiation for such purposes allows solving the given objectives.

The issues of applying SHF radiation in technical systems, including transport, are covered in [8–11]. Analysis of available publications shows that as for the efficiency of warming-up and, to a greater extent, maintaining the optimum temperature of TF the given method is the most competitive. However, up to now the following sequence has been poorly studied: treatment of the TF (fuel – regular, gas; oil; cooling liquid) – change of its physical and chemical properties – influence of the changed TF on the parameters of the internal combustion engine (ICE) working process (starting qualities, fuel consumption, environmental parameters). Meanwhile, this process particularly presents a wide interest, since the influence on TF at molecular and atomic levels can lead to any sort of changes in their physical and chemical properties. Thus, research in the given area will be the basis for identification of all qualitative changes in TF and evaluation of positive and negative sides of the given effect on TF.

The research focuses on developing a technique and device to improve vehicle efficiency by means of preheating that would not have thermal inertia, with the liquid heated instantaneously in the entire volume using less energy compared to conventional heating methods. Preheating by SHF radiation satisfies these conditions. After being treated with the SHF electromagnetic field, dielectric molecules are polarized and perform mechanical oscillations producing the energy which, due to intermolecular friction, is converted into heat and heats the dielectric. Heating takes place in the SHF electromagnetic field within the range of frequencies determined by the process requirements based on the volume of TF, its physical and chemical properties, initial and final heating temperature and other factors. The frequency of electromagnetic oscillations and, as its function, the wave length of the electromagnetic field is selected (calculated, determined experimentally) so as to be smaller than the geometric size of a compartment that contains TF. In this case, the electromagnetic field inside the compartment has a wave nature and repeatedly passes through the liquid, being reflected from the walls. However, SHF treatment of TF will inevitably lead to a change in its physical and chemical properties, since exposure occurs at the atomic level, which will affect its performance in the ICE. In this regard, we need to understand what changes occur in a particular fluid (changes in oil viscosity, its detergent power, flash point, chilling point, octane number, etc.).

The next step of the research must involve evaluating the ICE performance on the modified fuels, which results in development of the recommendations on how to treat a certain fluid or whether such treatment is allowable at all. In case of a complex positive result we need to develop comprehensive approaches to creating a system of preheating and maintaining the optimum temperature of TF at the given level on the example of preheating gaseous fuel with SHF radiation when it enters the engine. For the machinery to operate efficiently it is necessary to maintain its TF temperature at the optimum level. The optimum TF temperature ensures the lowest wear of units and aggregates, reduces fuel consumption, decreases exhaust gas emissions, etc. The optimum oil temperature (engine, reduction gearbox) improves lubrication of friction members by improving oil penetration into a gap between contacting surfaces. The optimum hydraulic oil temperature (hydraulic system, hydraulic cylinders, shock absorbers) reduces the risk of failure of units operating under high pressure due to the optimum oil viscosity. The optimum temperature of the cooling liquid in the internal combustion engine provides the best conditions for carburation and combustion of fuel in the cylinders of the engine by improving fuel-air mixing process. The optimum fuel temperature improves carburation, increases the cycle temperature and, as a result, fuel combustion efficiency, and reduces fuel consumption and exhaust gas emissions at idle speed. The optimum washing fluid temperature ensures the washer efficiency and reduces the risk of accidents caused by a contaminated windshield or headlights of the vehicle. To be able to achieve the optimum TF temperature a unit, aggregate or system should be warmed up to this state. The problem of warming up the vehicle engine is particularly urgent at low air temperatures. This problem is related to a complex start of the power plant, therefore, additional means of starting and warming the engine together with the systems, units and aggregates up to the operating temperature are becoming more commonly used. Currently used preheaters operate on a general principle of heating the liquid by heat transfer from the heated body (electric heater, exhaust gas heat exchanger, cooling liquid heat exchanger). This principle has a long response time and does not allow the given fluid temperature to be maintained. Devices that provide this method, as mentioned earlier, have a greater metal content, are not reliable and require additional costs to ensure their functionality. Thus, it is essential to develop a method and device that will provide rapid heating of fluid and maintain the fluid temperature at a certain optimum level.

2.2 Motor fuel adaptation to low-temperature conditions

In Russia the actual quality of motor fuels produced by the oil refining industry significantly lags behind modern European standards. A comprehensive solution to the problem requires a drastic change in the ratio of refining processes towards developing modern processes of improving fuel quality, which is not possible without multi-billion investments [12]. Different aspects of improving the quality of motor fuels by refining methods are presented in various studies [12–15].

In view of insufficient volume of winter diesel fuels manufactured in Russia there is an enormous over-expenditure of fuel, due to the unavoidable use of summer diesel fuels in winter. An improvement of the low-temperature properties of diesel fuels, a decrease in cloud point, pour point and cold filter plugging point in particular, will considerably increase the efficiency of diesel fuel at low temperatures. The oil refining industry in Russia is limited in its ability to considerably and quickly change the ratio of the produced summer, winter and Arctic grades of diesel fuels. That is why the most realistic method of fuel adaptation to low-temperature operating conditions is the application of additives improving the low-temperature properties of fuel. Thus, great volumes of diesel fuel consumption at significant concentration of the known depressants required for obtaining a significant decrease in pour point and cold filter plugging point considerably limit the application of these additives in diesel fuels. It is necessary to develop an additive that will have a significant impact when its concentration in the fuel is very low. Thus, the application of a multifunctional additive that improves the properties of diesel fuels and characteristics of engines is the most perspective.

It should be noted that improving motor fuel properties by means of additives [16–20] has a great performance potential, which is demonstrated by a considerable range of additives and numerous patents on the global market. Certain properties of motor fuels (lubricating, detergent) can only be modified with the help of additives, and are not provided by changing the refining industry technology.

3 Results and discussion

3.1 Influence of SHF radiation on physical and chemical parameters of diesel fuels

The first stage of the experimental research was to determine the changes in physical and chemical parameters of diesel fuel after exposure to SHF radiation.

Diesel fuel (winter grade) complying with the requirements of standards was divided into 100-ml probes. Each probe was exposed to SHF radiation for a different period of time. In total, 20 probes of 100 ml each were tested. The total amount of fuel obtained during one investigation – 2 liters. Table 1 provides the initial data.

Further on, the fuel was studied by assessing the following parameters: kinematic viscosity, density, flash point in a closed cup.

Table 1. Initial data

Number of probe	Exposure time, sec	Magnetron capacity, W	Frequency, GHz
1	0	850	2.4
2	180	850	2.4
3	300	850	2.4
4	480	850	2.4

Kinematic viscosity was studied according to GOST 33-2000 «Petroleum products. Transparent and opaque liquids. Determination of kinematic viscosity and calculation of dynamic viscosity». The standard determines a method for determination of kinematic viscosity of liquid petroleum products, transparent and opaque liquids by measuring the time it takes for a certain volume of liquid to flow under gravity through a calibrated glass capillary viscometer.

Density was studied according to GOST 3900-85 «Petroleum and petroleum products. Methods for determination of density». The method is used to determine density of petroleum and petroleum products by an oil hydrometer. Essentially, the method involves submerging the oil hydrometer in the test product, taking readings on the hydrometer scale at a test temperature and converting the results into density at a temperature of 20°C.

Flash point in a closed cup was studied according to GOST 6356-75 «Petroleum products. Method of test for flash point in a closed cup». It establishes a method for determination of flash point in a closed cup. Essentially, the method involves determining the lowest temperature at which under test conditions above the combustible substance's surface a mixture of fumes and vapors with air is formed. It is capable of flashing in the air from a source of ignition, but their rate of formation is still insufficient for the subsequent combustion. For this purpose, the test product is heated in a closed cup at a constant rate with continuous stirring and tested for flash at certain temperature intervals.

The study results are given in Table 2.

Table 2. Results of studying the impact of SHF radiation on physical and chemical parameters of diesel fuel

№	Test parameter	Probe 1	Probe 2	Probe 3	Probe 4
1	Kinematic viscosity, mm^2/s	3.21	3.16	3.13	3.04
2	Density, kg/m^3	824	821	820	815

3	Flash point in a closed cup, °C	36	33	30	30

The results lead to the following conclusions.

Exposure of fuel to SHF radiation reduces the content of both aromatic and naphthenic and heavy paraffinic hydrocarbons. The authors can also assume that viscosity decreased due to changes in the molecular structure. There was a decrease in the number of rings in the molecules of naphthenic and aromatic hydrocarbons; the hydrocarbon chain of paraffinic hydrocarbons and the length of a radical of naphthenic and aromatic hydrocarbons became smaller. All this indicates the decomposition of high molecular weight hydrocarbons. This is also supported by the results of determining the flash point in a closed cup. The lower the flashpoint, the more light hydrocarbons (low molecular weight paraffins) it contains.

3.2 Influence of SHF radiation on the engine output parameters

The next step of the research involved evaluating the ICE performance on fuels treated with SHF radiation. Tests were conducted on a diesel generator SDMO T9KM with a diesel engine Mitsubishi S3L2 SD. Table 3 presents main technical characteristics of the diesel generator. The test parameters were: fuel consumption per hour, kg/h; percentage of NO_x in exhaust gases, % vol.; smoke from the exhaust, %.

Factors considered during tests: fuel temperature, + 20 °C; air temperature, + 20 °C; atmospheric pressure, 750 mm of mercury; engine load, 70% of the nominal; rotational rate, 1500 min^{-1}. The values of fuel consumption were recordered from the dashboard of the diesel generator.

The NO_x percentage in the exhaust gases was measured using a gas analyzer GIAM-29M-3, which technical characteristics are shown in Table 4.

The smoke from the exhaust was determined using an opacimeter META-01MP 0.1 LTK, which technical characteristics are shown in Table 5. The test results are shown in Table 6.

Table 3. Technical characteristics of the diesel generator

Characteristics	**Values**
Capacity, kW	8.6
Cooling type	liquid
Nominal speed, min^{-1}	1500
Make of AC generator	Mecc Alte ECP 3-3L/4
Fuel consumption l/hour (at 70% load)	2.4
Noise level	60.4 dB

Table 4. Technical characteristics of gas analyzer GIAM-29M-3

Characteristics	**Values**
Type of gas analyzer	portable
Way of probe sampling	forced
Principle of action	electrochemical
NO measuring range, % vol.	0–0.5
Limit of the tolerable basic reduced error, NO %:	10
Ambient temperature, °C	from 0 to +45

The diesel fuel treatment led to a reduction in the smoke from the exhaust to 15%. This can be explained by a decrease in the number of areas with a lack of oxygen because of improved mixture formation that was caused by a decrease in the viscosity and density of the fuel.

There is a reduction in the nitrogen oxide emissions to 30%, which is due, among other things, to a reduction of the fuel flash point in a closed cup. This corresponds to a decrease in the self-ignition temperature of the fuel and an improvement of combustion at a lower fuel density. The fuel burns more uniformly, without temperature peaks.

The reduced fuel consumption down to 7% may indicate an improvement in the process of mixture formation and combustion, due to an increased amount of light hydrocarbon fractions.

Table 5. Technical characteristics of the opacimeter META-01MP 0.1 LTK

Characteristics	**Values**
Type of device	portable
Smoke measurement range in terms of absorption coefficient K, not less than, m^{-1}	0-∞
Smoke measurement range in terms of attenuation coefficient N, %	0–100
Absolute limit of error at absorption coefficient of 1.6 – 1.8 m^{-1} not greater than, m^{-1}	±0.05
Nominal scale point of the least category for absorption coeff., m^{-1}	0.01
Nominal scale point of the least category for attenuation coeff., %	0.1
Photometric base, m.	0.1
Automatic adjustment of smoke readings by the temperature of exhaust gases	enabled
Automatic zero adjustment	enabled

Table 6. Results of studying the impact of SHF radiation on the engine output parameters

№	**Test parameter**	**Probe 1**	**Probe 2**	**Probe 3**	**Probe 4**
1	Fuel consumption per hour, kg/h	2.38	2.31	2.24	2.21
2	Percentage of NO_x in exhaust gases, %	0.03	0.03	0.026	0.023
3	Smoke from the exhaust, %	60	58	55	52

3.3 Improvement of low-temperature properties of diesel fuels using additives

The impact of depressants is the result of adsorption of additive's molecules on the surface of a paraffin crystal that prevents its further growth. Thus, smaller crystals are formed and their coalescence, leading to the formation of three-dimensional paraffin structure causing oil solidification, is less likely to occur. For diesel engine operation the most important indicator of low-temperature properties is a cold filter plugging point; however, a large difference between the cold filter plugging point and cloud point is ineffective, as small crystals passing through the filter are deposited, disrupting fuel supply.

The influence of depressants on the cloud point is possible if the depressant molecule when reacting with alkane molecules with a high melting point prevents the formation of paraffin crystal nuclei.

The acting mechanism of depressants makes a lot of substances depressant-active if their molecules contain a fragment with alkane properties (long unbranched alkyl radical) and a polar group or a group radically different in structure: alkylnaphthalene, soaps of polyvalent metals, relatively low molecular weight polymethacrylates (molecular weight ~ 17000) containing n-alkyl radicals C_{12}÷C_{18}, low molecular weight copolymer of ethylene and vinyl acetate and others. One of the most effective depressant additives, named Keroflux 5486, is manufactured by BASF. At concentration 0.05% it decreases the pour point of summer grades of diesel fuels by 15–18° and the cold filter plugging point by 14–16°.

A multifunctional additive $[RCOO]_2Me$, where $R=C_9H_{19}÷C_{15}H_{31}$, was developed, providing a comprehensive positive effect on the properties of the fuels used [18]. Investigations of the influence of the additive on the low-temperature properties of diesel fuels have been conducted.

At extremely low concentrations in diesel fuel $[RCOO]_2Me$ significantly (by 8–9°) reduces the cold filter plugging point (Table 7).

Table 7. Impact of $[RCOO]_2Me$ at concentrations 27.75 mg/kg/18.5 mg/kg in diesel fuel on the low-temperature properties of fuels

Diesel fuel	Reduction in temperature, °C		
	cloud point	pour point	cold filter plugging point
Sample I (pour point –7°)	–4/–3	–9/–8	–8/–9
Sample II (pour point –14°)	0/0	–2 /0	–9 /–8

The obtained results based on existing ideas about the low-temperature properties of fuels are explained by the increase in dispersion of the released paraffin crystals due to blocking the crystal growth by adsorption of the additive's surfactant molecules on their surface.

The additive significantly – by up to 10% – reduces the diesel fuel consumption at low ambient temperatures, both by improving the low-temperature properties of fuels, and as a result of eliminating carbonization in the engine [19]. The additive also has anti-wear properties, improving the lubricating ability of diesel fuel up to 25%, which significantly reduces the wear of plunger pairs in high-pressure fuel pumps of diesel engines. At the same time, the environmental performance of diesel engines significantly improves: CO emissions decrease by 15–17%, nitrogen oxide emissions – by 20–22%, hydrocarbons emissions – by 35–37%, soot emissions – by up to 40–50%. It should be noted that the use of additives does not result in the formation of any additional toxic substances in exhaust gases that have not been previously observed.

4 Conclusions

Treatment of diesel fuel with SHF radiation improves its environmental performance and reduces fuel consumption.

The results obtained are the basis for the development of a neural network control system of maintaining the optimum temperature of diesel fuel in a vehicle for the «smart city» system that improves its efficiency and environmental safety due to neural network control of units and aggregates. Neural network control will enable the improvement of the adaptability of vehicles to variable low-temperature operating conditions according to fuel consumption and environmental indicators.

Application of the proposed multifunctional additive improves the low-temperature properties of diesel fuel and has a complex positive impact on the diesel fuel properties and the state of engines. The additive enables a significant decrease in fuel consumption under low-temperature operating conditions and improves the environmental safety of vehicle operation.

Acknowledgement

This research was supported by Act 211 Government of the Russian Federation, contract № 02.A03.21.0006.

References

1. N. Karnaukhov, L. Reznik, V. Holyavko, Adv. Transport, **6**, 439–443 (2000)
2. I. Anisimov, A. Ivanov, E. Chikishev, D. Chainikov, L. Reznik, WIT Trans. Ecol. Envir., **190** (**1**), 685–695 (2014)

3. J.A. Ertman, S.A. Ertman, I.A. Anisimov, D.A. Chainikov, Y.M. Chikishev, Research Journal of Pharmaceutical, Biological and Chemical Sciences, **6** (**1**), 1761–1770 (2015)
4. B. Torregrosa-Jaime, F. Bjurling, J.M. Corberan, F. Di Sciullo, J. Paya, Appl. Therm. Eng., **75**, 45–53 (2015)
5. C. Bauer, J. Hofer, H.J. Althaus, A. Del Duce, A. Simons, Appl. Energ., **157**, 871–883 (2015)
6. A.P. Roskilly, R. Palacin, J. Yan, Appl. Energ., **157**, 563–566 (2015)
7. E. Magaril, L. Reznik, *Urban Transport and the Environment II,* 149–154 (1996)
8. M. Novotný, K. Šuhajda, J. Sobotka et al., Wood Res-Slovakia, **59** (**3**), 389–400 (2014)
9. D. Nowak, M. Stachowicz, K. Granat, M. Pigiel, Metalurgija, **52** (**2**), 211–214 (2013)
10. K. Granat, B. Opyd, D. Nowak, M. Stachowicz, G. Jaworski, Arch. Metall. Mater., **58** (**3**), 919–922 (2013)
11. A. Arshanitsa, Y. Akishin, E. Zile et al., Renew. Energ., **91**, 386–396 (2016)
12. E. Magaril, WIT Trans. Ecol. Envir., **190** (**2**), 821–832 (2014)
13. R.R. Shiriyazdanov, U.S. Rysaev, E.A. Nikolaev et al., Chem. Tech. Fuels Oil+, **44** (**5**), 295–297 (2008)
14. P.R. Robinson, in: *Practical Advances in Petroleum Processing*, **1**, 1–78 (New York, 2006)
15. A. Stanislaus, A. Marafi, M.S. Rana, Catal. Today, **153** (**1–2**), 1–68 (2010)
16. J. Bennett, in: *Alternative fuels and advanced vehicle technologies for improved environmental performance. Towards zero carbon transportation,* 165–194 (2014)
17. A.M. Danilov, Chem. Tech. Fuels Oil+, **47** (**6**), 470–484 (2012)
18. E. Magaril, WIT Trans. Ecol. Envir., **147**, 373–384 (2011)
19. E. Magaril, International Journal of Sustainable Development and Planning, **8** (**2**), 231–245 (2013)
20. Srivastava, J. Hancsók, *Fuels and fuel-additives* (New Jersey, 2014)

E3S Web of Conferences **6**, 02004 (2016) DOI: 10.1051/e3sconf/20160602004
ICSC 2016

Ensuring sustainability of the city transportation system: problems and solutions (ICSC)

Irina Makarova[1], *Rifat* Khabibullin[1], *Ksenia* Shubenkova[1a], and *Aleksey* Boyko[1]

[1]"Service of Transport Systems" Department, Kazan federal university, 423810, Naberezhnye Chelny, Russia

Abstract. Ways to increase sustainability of the city transportation system and, particularly, measures to promote safe public transport and non-motorized means of transport are considered in the article. Analysis of the existing positive experience shows that complex decisions for a sustainable development of the city transportation system are necessary. Technical and organizational ways to increase the transportation system's sustainability and safety are studied. The results of solution of separate tasks, aimed at completing the goal, are presented: defining transport preferences of the citizens of Naberezhnye Chelny, perspective model of bus route network, a model of an adaptive smart-bicycle. It is shown that the proposed solutions for strategic and operational management will help to enhance efficiency and safety of transportation system.

1 Introduction

During the debates about sustainable development the participants of the United Nations Conference on sustainable development (Conference "Rio+20" [1]) noted that transportation and mobility are central to sustainable development as means to improve social equity, health, resilience of cities, urban-rural linkages and productivity of rural areas. They recognized the need to promote an integrated approach to policymaking at the national, regional and local levels for transport services and systems to promote sustainable development. Sustainable transport systems development is one of the Global Goals for Sustainable Development [2] and is the priority area of work of the Pan-European Program on Transport, Environment and Health (the PEP). In the Paris Declaration in addition to the four priority goals of the PEP the new priority goal was adopted. This goal is to integrate transport, health and environmental objectives into urban and spatial planning policies by developing capacities and frameworks for integrated urban and spatial planning in order to reduce the impact of transport on health, the environment and land use, increase energy efficiency and support green and healthy mobility and transport as well as sustainable livelihoods [3]. According to the Global Status Report on Road Safety 2015 [4], moving towards more sustainable modes of transport (such as cycling and public transport) has positive effects if associated road safety impacts have been well managed.

[a] Corresponding author: ksenia.shubenkova@gmail.com

These include increased physical activity, reduced emissions and noise levels, reduced congestion and more pleasant cities.

In Russia, the safety of transport systems is the issue of the day, because number of highway fatalities up 10 thousand vehicles is higher by 10–15 times, and up 100 thousand people is 2–4 times higher than in other countries where motorization rate is higher and the network capacity is the same [5]. The reasons for the increasing number of road accidents lie not only in developmental lag of transport infrastructure, but also have organizational nature. Thus, Moscow roads have been designed for a 4 million vehicles maximum, while the number of registered vehicles today is about 4.5 million. This is the reason of the fact that the average vehicle speed today is 7 ÷ 11 km/h and the drivers have to spend a lot of time in traffic jams. Even the day-time lorry traffic ban in the city center could not significantly improve the situation [6]. Another problem faced by an ordinary vehicle driver is the lack of parking space in the city center. In other major Russian cities, the situation is not much better, and even youngish cities where are wide avenues are familiar with the problem of traffic jams and lack of parking space.

2 Ensuring sustainability of the city transportation system

2.1 Ways to increase sustainability of the city transportation system

Sustainability of the city transport system is usually means the effective organization that is characterized by the following: a) transport infrastructure provides safe and comfortable travel to all road users: to the drivers of private vehicles, to the passengers of public transport, to cyclists and pedestrians, regardless of their physical disabilities; b) public transport provides affordable (financially and physically) services to all citizens and it is an acceptable alternative to private vehicles when their using leads to overloading of the road network; c) the way of traffic organization ensures the most efficient use of the existing road network, reduction of traffic jams and negative transport-related impact on the environment [7]. One of the most important factors that confirm the effectiveness of the organization of the transport system is existence of strategic planning. Methods to ensure sustainability that are developed within the framework of strategic planning are based on finding a balance between transport demand and supply of the most effective and safe modes of transport (Figure 1).

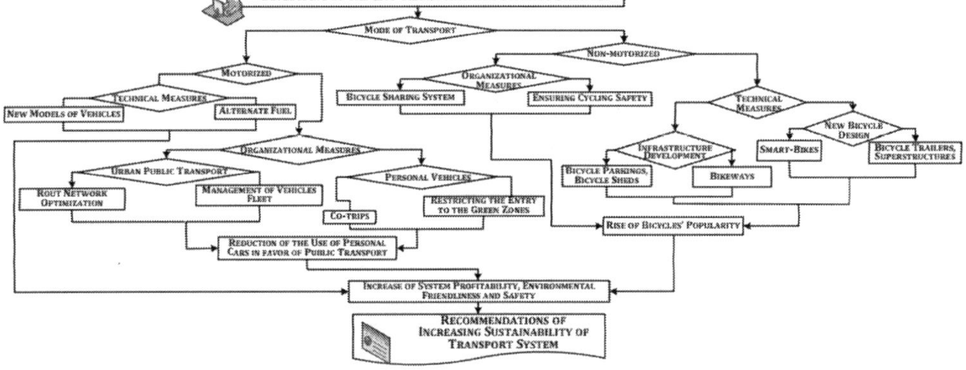

Fig. 1. Methods of increasing sustainability of the city transportation system.

Moving towards more sustainable modes of transport has positive effects if the associated road safety impacts have been well managed. Aside from reducing road traffic injuries, there are positive health benefits that are associated with increased physical activity, reduced pollution, noise levels and greenhouse gas emissions, reduced congestion and more pleasant cities. Moreover, measures to promote safe public transport and non-motorized means of transport are also in line with other global moves to fight obesity and reduce noncommunicable diseases (such as heart disease, diabetes) [8].

2.2 Positive examples of increasing sustainability of the city transportation system

In the documents of the PEP it is noted that there are inspiring examples of improving the quality of the urban environment and supporting a modal shift towards cycling and walking in combination with public transport. These positive changes include the investments of cities such as Paris and Barcelona in cycling infrastructure and city bikes, the introduction of congestion charges in London, Stockholm and other cities and interventions made in New York to "pedestrianize" highly congested areas and turn them into parks [9]. It is also noted that electric mobility is becoming more affordable and widespread. Many corporate vehicle fleets and vehicle-sharing schemes across the pan-European region rely on electric and/or hybrid vehicles and the introduction of electric bicycles has made it possible to use cycling not only for health and leisure, but as a viable means of transport. Among the factors constraining the development of alternative ways of movement is the fact that only 68 countries have national or subnational policies to promote walking and cycling, and just 79 countries have policies that protect pedestrians and cyclists by separating them from motorized and high-speed traffic [10].

In Germany as part of the «50 by 50» campaign [11] some communities have started moving towards the «carfree city», for example, one such community is Vauban (Germany). This approach is ensured by high-quality public transport, walking and cycling facilities, as well as limited and costly parking [8]. The government of the neighbor city Freiburg in order to achieve objectives of transport-related environmental pollution abatement is constantly investing in the non-motorized transport development. Last years about 6000 bicycle parkings (including tandems and bicycles with a trailer) have been built there. The main purpose of this action is to make bicycles popular not only among students but also among businessmen, women and children. In order to incorporate cycling into a united city transportation system bicycle parkings have been built also at several selected bus stops of the public transport [12].

The bicycle sharing system was implemented in the Republic of Korea in the city of Changwon, where old technology (the bicycle) has been matched up with the new technology (smartcards), so that bikes can be used on demand, either free for an initial period or for a reasonable charge. Bicycle sharing system is also implemented in Hangzhou (China), where the popularity of this mode of transport is also provided by the widespread introduction of electric bicycles [8]. Today, road users of many European cities (such as Oulu, Amsterdam or Copenhagen) at two-thirds consist of cyclists. However, not everyone can ride a bicycle every day, especially long-distance cycling. That is why bicycles should be considered not as a competitor, but rather as an addition to other modes of transport. Thus, the use of personal cars for daily trips may be reduced providing: the right transport links between all parts of the are established by public transport; sufficient reasons for not owning a vehicle (e.g. limited and costly parking); local walking and cycling facilities and services.

In most cases of successful implementation, however, it is not a single policy that has been introduced – but rather a package of measures. Such 'packages' are more likely to gain public acceptance, and they allow a mixture of policies that may be seen as disadvantageous to individual users, but promote overall welfare gains to society [8].

2.3 Improving the quality of public transport governance

The shift from personal cars to public transport without sacrificing mobility of the population can be ensured by increasing the safety and comfort of its services [13] that is considerably has a direct link with improving the quality of public transport governance. The operational management of large systems, which also include transportation system, decision support systems (DSS) are used. Intelligent hearts of such systems can be various means that provide analysis of the operational information about system parameters, such as: parameters of traffic flow, road accidents and other emergency situations. Such systems may include sets of canned solutions for different factor combinations that facilitate, if necessary, a reasonable selection of the best managerial solution.

These decisions can be laid at the strategic planning stage. The management quality indicators are increasing of safety, reliability and sustainability of the system. The correctness of such a system's operation is largely determined by the quality of the initial information and the adequacy of its analyzing and processing methods. For analysis and development of recommendations in the DSS information search, data mining, knowledge retrieval in databases, reasoning based on precedents, simulation modeling, evolutionary computation and genetic algorithms, neural networks, situational analysis, cognitive modeling, etc are used. There are three main parts in the DSS: 1) Data system for collection and storage of information received from internal and external sources, as a rule, it is a data store; 2) the system of dialogue that allows the user to specify what data to choose and how to handle them; 3) the system of models – ideas, algorithms and procedures that allow to process data and to analyze them. The user has the experience, knows the situation and follows certain considerations when fetching data. In the processing of data, use different procedures, from simple summing to statistical analysis and nonlinear optimization.

The intellectual heart of the designed system is a transport model, which is a software package, consisting of a Network Model, Travel Demand Model and Impact Model. The Network Model is a representation of the road network and in that way means a city transport supply. The Travel Demand Model contains the travel demand data (cost matrix, origin-destination matrices, etc.), and describes the transport needs of the population with a standard four-stage model usage. Impact Models allow to evaluate comprehensively each option of the traffic organization and to choose the best from the point of view of increasing the efficiency of use of the vehicle fleet of transport enterprises, and with a view of minimizing the negative impact on the environment without compromising it's mobility [14].

Origin-destination matrices creating and the traffic flows distribution allows using Network Model and Travel Demand Model that take into account parameters of the real system. It is necessary to have qualitative operational information on these current system parameters values for the correct model simulation. Realization of the concept of "Safe City" (in Russian cities) contributes to improving the quality of information through the paragraph "Transport Security" [15] where the objectives for road safety are formulated. Thus, in Naberezhnye Chelny to the end of 2016 1020 video cameras will be installed on dwellings, social and cultural objects and transport infrastructure objects (roadways, public transport stops and passenger compartments, etc.). This will allow collecting operational information and adjust the operational management based on data analysis.

Simulation models designed just as for the whole city, so for the areas with critical values of traffic parameters, enable to choose the best options for different values of the initial parameters. As a result of the simulation traffic loads are calculated that helps to identify problem areas that should be optimized. After the model verification and validation it can be used to correct city route network in order to avoid buses passing through the overloaded road network areas. After the route network correction it is the structure of the vehicle fleet on the bus routes that is optimized according to their operating characteristics. And then the traffic schedule is formed, taking into account the maximum allowable traffic intervals. Economic effect of DSS application is determined by the reduction of transportation costs by optimizing routes and increasing efficiency of the vehicle fleet. Environmental effect is to reduce the negative impact on the environment and can be achieved through rational redistribution of traffic flows on the city road network. Social effect will be the reduction of the risk of disease due to urban air quality as well as in decreasing delays and the fatigue of the population in the process of reaching destinations.

2.4 Development of a sustainable system of non-motorized transport

Bicycling makes efficient use of roadway capacity and reduces congestion. It is a low-polluting and a low-cost transportation alternative and can be an important means for getting to destinations that are not serviced by transit [16]. The advantages of cycling include cheap infrastructure requirements and improvements in public health. Bicycle pathways, lanes and parking require less space than their automobile counterparts. Cycling has direct health benefits. It is an aerobic exercise that can minimize the risk of muscle and ligament injury, lower blood pressure and reduce the risk of heart disease) [17]. Moreover, in urban areas, cycling can sometimes prove to be faster than other transport modes and also allows cyclists to avoid traffic jams. But at the same time cycling has a number of disadvantages, however, including a greater physical effort, the difficulty of carrying loads while cycling, being at the mercy of the weather, and, outside urban areas, travelling more slowly than motorized transport. Factors such as physical effort and speed also limit the distance that a cyclist can travel [18]. These activities may promote cycling: the design of cycle paths, taking into account such factors as the terrain; the elaboration of new bicycle designs, such as smart-bike, electric bike, cargo bike, bike for passengers transportatio (children, the disabled people, the old people), etc.; integration of bicycles to public transport infrastructure (bike parkings at public transport stops, the possibility to take the bike in buses or trams); development of cycling infrastructure: safe bike paths with rain, wind and sun protection, secure bicycle parkings at all destinations, bicycle sharing system, etc.

The attractiveness of cycling may be increased by the expansion of the model line-up of bicycles for different population groups and various use cases. One of the solutions could be a smart bike usage. Currently electric bicycles have found wide application. They are generally different from ordinary bicycle because of three additional components presence such as an electric motor, a storage battery and a battery controller. In contrast to the electric scooter or motorcycle, electric bike may be driven by pedals and it's operation and maintenance is a little more complicated than an ordinary bicycle handling. Despite of electric drive presence electric bike is used approximately the same as an ordinary bicycle and in most countries does not require the driving license or license plate presence. Electric bicycle is suitable as a vehicle for a wide range of people with the different level of abilities, as it is easy to dose physical training. There is a number of disadvantages of electric bicycle that, makes it difficult to use. They are: significant weight (from 20 to 50 kg or more) and the corresponding inertia; lack of power reserve on the drive (rarely more than 25–50 km); long battery charging (usually at least 2–6 hours); short service life of lead-acid and lithium-Ion storage batteries; the high cost of the final product and its use compared with an ordinary bicycle cost and use (from 2 to 10 times).

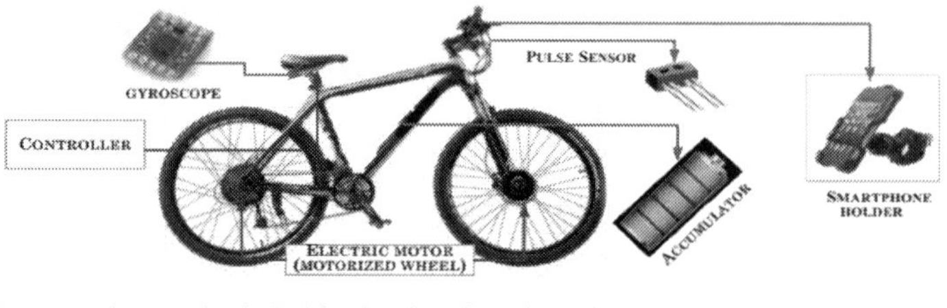

Fig. 2. The elements that are included in the developed module.

Electric bicycles are controlled by bicycle computer (controller), which is supposed to: supply amperage from the battery to the electric motor in accordance with the user's settings; show residual battery charge on the indicator; determine the rotation / stop of pedals; limit the maximum speed of the bicycle movement in order to save energy;keep constant speed (cruise control); charge the battery while braking. At the same time there is a variety of stationary bicycles that are belong to the group of cardiovascular machines which are equipped to control the physical condition of a user. At the same time the main indicator to diagnose critical state is a pulse rate. If to equip the bicycle with the universal module, which includes a pulse sensor, a controller and other components that are shown in Figure 2, and to manage it in accordance with the selected program installed on smartphone it will help to increase the attractiveness of cycling among untrained population. Application of the developed module will enhance the attractiveness of the bicycle due to the fact that owners of such vehicles may have insufficient training and at the same time the overload risk will reduce.

3 Application of imitation models to choose the best way of increasing sustainability of the city transportation system

Addressing the issues of urban mobility ensuring that is directly related to the city transportation sustainability should be based on and taking into account the preferences of the population. Only in this case the projects will be successful, and it ensures the effectiveness of investments to transport system infrastructure. To justify the choice of the best options there are two important aspects in strategic planning. They are the adequacy and quality of the initial information on the current state of the transport system as well as the preferences of the population that is necessary for predictive models design. The best tool for the comparison of options is a simulation. One of the most objective methods to study the transport preferences of the population is a questionnaire survey, that allows to predict the most likely options of the transportation system development. These surveys should be carried out in proper structuring and data processing and also with an allowance for risks.

To find out what transport modes population of Naberezhnye Chelny prefer and what are the transport problems from the consumers' point of view the questionnaire has been developed. 953 respondents, constituting the various target groups, took part in the current survey (Table 1). To analyse possible directions of Naberezhnye Chelny transportation system's development a comparison of transport mobility indicators of its population and the population of Kopengagen, that is similar in size [19], was made (Table 2).

Table 1. The results of the sampling survey of population of Naberezhnye Chelny.

Indicator	Students	Workers	Retiree	Other category	TOTAL
number of respondents	624	299	16	14	953

number of trips to work or education by public transport	313	109	-	-	422
number of trips to work or education by bicycles	50	7	-	-	57
number of trips to work or education by cars	163	133	-	-	296
number of trips to work or education by foot	98	50	-	-	148
the number of drivers who are ready to transfer to bicycles, if there are:					
• bikeways	127	56	0	0	183
• bicycles parkings	129	46	0	0	175
• bike hire system	75	28	0	0	103
• the possibility to take the bike in buses or trams	76	21	0	0	97
• E-bikes	78	22	0	0	100
the number of drivers who are ready to transfer to public transport, if there are:					
• busstops can be reached by foot in less than 5 minutes	85	38	1	0	123
• bus waiting time is 5 minutes or less	89	47	1	1	136
• direct connection, no transfers	73	45	1	1	118
• vehicles are clean and comfortable	94	40	1	0	134
• place to sit	78	37	1	0	116
travel modes preferences:					
• buses	127	88	6	6	227
• minibuses	371	115	3	6	495
• trams	134	59	7	1	201
number of bikes in the personal property	313	86	2	6	407

Priority strategy of politicians in Copenhagen is development of bicycles infrastructure as a way to create more friendly city living condition. The conditions for the implementation of a "green transport" strategy are also created in Tatarstan as well as in a number of Russian cities. Moving towards more sustainable modes of transport (for example, gas and hybrid buses, popularization of cycling) is realizing in order to support this strategy. Any changes in the city transport infrastructure should take into account a great amount of factors, as well as the preferences of the population of the city. The results of the survey conducted in Naberezhnye Chelny have shown that reducing the number of trips by vehicle in favor of the public transport can be achieved by route network improving, as well as the optimization of structure of the vehicle fleet on the bus routes.

Table 2. The comparison of transport mobility indicators.

Indicator	Kopengagen	Naberezhnye Chelny
number of inhabitants	550000	522000
city area, km^2	88,25	171,03
number of vehicles in the personal property	125	147
number of bikes in the personal property	650 000	n/a
number of cars per capita	0,23	0,28
average number of bikes per capita	1,2	0,43
bikeways length, km	350	110
bikeways length per square of the city, км/км^2	3,96	0,64
trips to work or education by public transport, %	32	44
trips to work or education by bicycles, %	36	6
trips to work or education by personal vehicles, %	26	31
motorized and non-motorized transport integration:		
• the possibility to take the bike in buses or trams	+	-
• bike parkings at public transport stops	+	-
secure bicycle parkings at all destinations	+	-
public transport operates day and night	+	-

availability of shortened bus routes	+	-
availability of high speed bus routes	+	-
the presence of additional routes during peak hours	+	-

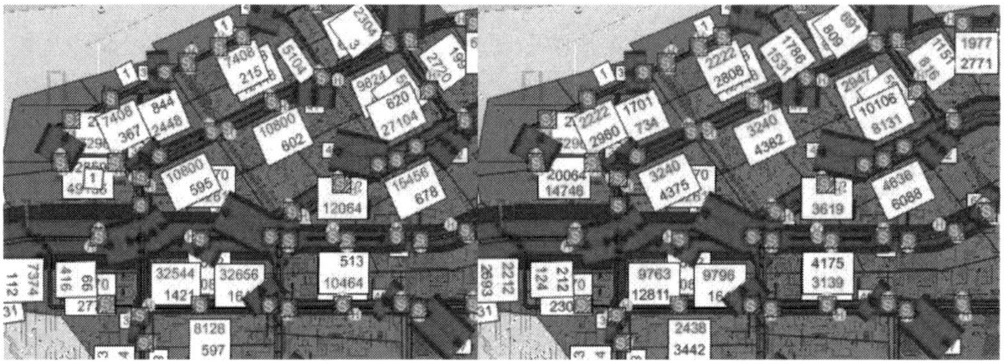

a) b)

Fig. 3. The traffic load distribution: a) existing passenger flow and route network; b) predicted passenger flow.

Firstly, predictive model of the new route network has been developed taking into account the following conditions: a) route network optimization should be carried out in such a way that, on the one hand, to eliminate (as far as possible) the imposition of routes, and on the other, to link all traffic areas of the city through direct routes; b) according to respondents, about 70% of drivers are willing to stop using private cars in favor of more sustainable modes of transport. That's why changes of modified trip distribution coefficients were introduced into the Travel Demand Model and the predicted values of traffic loads on the road network areas were calculated; c) while the vehicle fleet structure and the bus schedule optimization the predictive number of passengers should be taken into account. What's more it should be adjusted by changing traffic parameters. The suggested variants checking should be carried out with the help of simulation models. There is the road network area before and after the suggested changes in the Figure 3 (a, b). The red marks number of personal cars on the area, and blue marks the number of public transport passengers. Data on public transport demand allow to calculate the optimal number of different types of vehicles for each route. Thus, increasing the attractiveness of the public and non-motorized modes of transport will make the traffic load almost two times lower, and improve the ecological situation in the city.

4 Conclusion

Sustainable urban mobility requires high-performance public transport system and a combination of different modes of transportation. At the same time it has significant potential for the cycling development, because this clean and efficient kind of transportation contributes not only to the increasing of the city transportation system sustainability, but also to the economy's development. The experience of developed countries shows that in England bicycle accessories selling brings £ 853 mill. per year, 23,000 people employed in the cycling industry get £ 500 mill. per year and respectively pay £ 100 mill. of taxes. There are a lot of obstacles to the widespread use of the bicycles as an alternative travel mode in Russia. However, the development of an adaptive E-bike, bicycles infrastructure and it's integration to the public transport system will contribute to use of bicycle and public transport. City transportation system is a complex multi-parameter system, that's why every potential change in the traffic and passenger flows structure should be assessed comprehensively using DSS. The developed control system allows not only to analyze "what if...", but also evaluate each variant of arranging public transport's traffic and to choose the best both in terms of more efficient use of the vehicle fleet of transport enterprises and with a view to minimizing the negative impact on the environment while maintaining the mobility of the population. Thus, moving towards sustainable transport in Russian cities is possible today and depends on depends on the presence of supportive and nurturing governance, as well as sound institutional and regulatory structures.

References

1. The future we want. Rio de Janeiro: United Nations (2012)
2. Global status report on road safety 2015 (WHO Library Cataloguing-in-Publication Data, 2015)
3. Paris Declaration. City in motion: People first (Geneva: United Nations Publications, 2015)
4. E.B. Barbier, *A global green new deal: Rethinking the Economic Recovery* (Cambridge: Cambridge University Press 2010)
5. M.V. Zelencov, *Improving city traffic safety based on neural networks and discriminant analysis techniques* (Orel, 2010)
6. Changing urban traffic and the role of bicycles: Russian and International Experiences (Friedrich-Ebert-Stiftung, 2014)
7. The national concept of sustainable city transportation systems (2013)
8. Global Report on Human Settlements (UN-HABITAT, 2013)
9. Preparation of the Fourth High-level Meeting on Transport, Health and Environment (United Nations Publications, 2014)
10. Global status report on road safety 2013: supporting a decade of action (Geneva: WHO Library Cataloguing-in-Publication Data, 2013)
11. Share the Road Programme (UNEP, 2015)
12. A.V. Finogenov, *Urban development, Vision Expo*: Report (2010)
13. Green and healthy jobs in transport: launching a new Partnership under THE PEP (2011)
14. I.V. Makarova, R.G. Khabibullin, K.A. Shubenkova, World transport and technological machinery, **50**, 103–114 (2015)
15. The order of the Russian Federation "The Concept of developing the hardware and software complex "Safe City". Available at: www.rg.ru/2014/12/11/gorod-site-dok.html
16. S. Handy, Y. Xing, International Journal of Sustainable Transportation, **5**, 91–110 (2011)
17. S. Tsenkova, D. Mahalek, Urban, Planning and Transport Research, **2**, 126–146 (2014)
18. E. Heinen, B. van Wee, K. Maat, Transport Reviews, **30**, 59–96 (2010)
19. Lonely planet: Denmark. Available at: http://www.lonelyplanet.com/denmark/copenhagen

E3S Web of Conferences 6, 02005 (2016) DOI: 10.1051/e3sconf/20160602005
ICSC 2016

Universal computer vision system for monitoring the main parameters of wind turbines

Sergey Korzhavin[1a], *Sergey* Shcheklein[1], *Alexey* Zavyalov[1], *Konstantin* Purtov[2], and *Vladimir* Kublanov[2]

[1]Department of nuclear power plants and renewables, Ural ENIN, Ural Federal University, 620002, Yekaterinburg, Russia
[2]Scientific research medico-biological engineering center of high technologies, IRIT-RTF, Ural Federal University, 620002, Yekaterinburg, Russia

Abstract. The article presents universal autonomous system of computer vision to monitor the operation of wind turbines. The proposed system allows to estimate the rotational speed and the relative position deviation of the wind turbine. We present a universal method for determining the rotation of wind turbines of various shapes and structures. All obtained data are saved in the database. The presented method was tested at the Territory of Non-traditional Renewable Energy Sources of Ural Federal University Experimental wind turbines is produced by "Scientific and Production Association of automatics named after academician N.A. Semikhatov". Results show the efficiency of the proposed system and the ability to determine main parameters such as the rotational speed, accuracy and quickness of orientation. The proposed solution is to assume that, in most cases a rotating and central parts of the wind turbine can be allocated different color. The color change of wind blade should not affect the system performance.

1 Introduction

Wind turbines have become full-fledged participants in the process of electricity generation. Despite the huge variety, the manufacturers continue their modernization and developing new designs and principles. It is not always possible to evaluate all of testing parameters of new designs or existing wind turbines in different climatic conditions.

Main parameters of the wind turbines are the speed of rotation of the wind wheel, accuracy and quickness of orientation in a fishtail wind. Evaluation of deformation and vibrations of new designs are important characteristics at system calibration and installing.

The recent works show a lot of different ways for determining the parameters of wind turbines [1]. To measure the speed, orientation, vibration and deformations of existing wind wheels it is often necessary to redesign the original installation structure. Such changes require considerable work, change the weight distribution and a starting speed, and may

[a] Corresponding author: blyblk@gmail.com

affect the quality of the wind generator. Therefore, it is appropriate to develop a system that determines main parameters of wind turbines without changing the design.

One of the promising areas of measuring quantitative characteristics of the wind turbine are the computer vision systems [2, 3]. Such systems are currently being used in the wind turbine stations with pre-defined types of wind generators.

Such systems are difficult to design and are not universal. Determination of the rotational speed in such systems usually occurs by means of edge detection and tracking methods. Such methods do not work well in terms of variability background, lighting, and also it impose restriction: wind generator should be as "propeller-type", it not allowing to work with "screw type" wind turbines.

This article discusses the possibility of applying computer vision application to assess the performance of wind turbines. We focused on creating a simple and universal method for determining the main parameters of the wind turbine.

2 Material and methods

2.1 Experiment

This work is performed at the Territory of Non-traditional Renewable Energy Sources of Ural Federal University.

The installation system consists of several components: multiple cameras, "screw type" wind generator, storage database, and personal computer, which process and analyze data. The scheme of installation system is shown in Figure 1

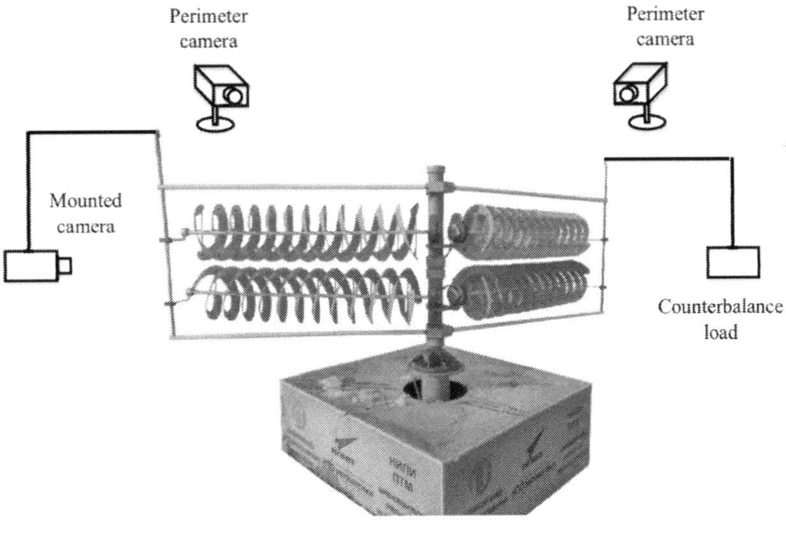

Fig. 1. The scheme of proposed system.

Experimental wind turbines are produced by "Scientific and Production Association of automatics named after academician N.A.Semikhatov". Each of them consists of two pairs of 'screw type' blades, which placed in pairs on the central part of the rotating wind turbine.

The proposed system consists of multiple network video cameras. The cameras allow to capture color frames approximately at 30 frames per second. They were deployed around the perimeter of the wind turbine, and one camera is rigidly fixed to the structure of the generator perpendicular to the axis of rotation. The presence of the camera was balanced on the other side by corresponding load.

The proposed computer vision software is installed on a personal computer that processes data from the camera in real time and transmits the processed data to the storage server [4].

The Figure 2 shows the examples of video images obtained by perimeter and mounted cameras.

a) b)

Fig. 2. The examples of video images obtained by a) perimeter network video camera, b) mounted network video camera attached to the wind turbine.

For evaluation of the wind turbine at the territory of laboratory we have the meteorological station, which allows to observe the speed and direction of the wind, with the ability of accumulation of these data.

2.2 Algorithm

The proposed algorithm is based on the assumptions that the central part and the rotating part of construction of the wind turbine can be allocated by individual colors from the background. It is desirable that the center of rotation and the blade rotation area were different colors.

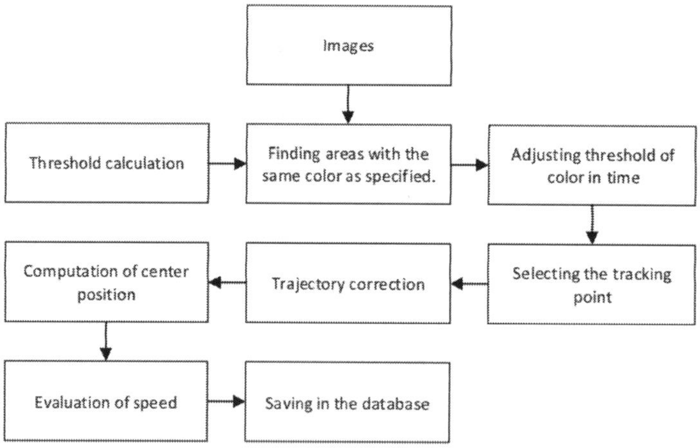

Fig. 3. Block diagram of the proposed algorithm.

The sequence of proposed algorithm:

1. Determination of thresholds for multiple color ranges and formats. Video frames are converted into YCbCr and HSV color formats and HSV to more accurately determine the color used. The YCbCr coordinates are calculated as follows:

$$\begin{aligned} Y &= 0.299 \cdot R + 0.578 \cdot G + 0.114 \cdot B \\ C_r &= (R - Y) \cdot 0.713 + 128 \\ C_b &= (B - Y) \cdot 0.564 + 128 \end{aligned} \tag{1}$$

To calculate the HSV coordinates R, G, B are converted to the floating-point format and scaled to fit the 0 to 1 range:

$$\begin{aligned} V &= \max(R, G, B) \\ S &= \begin{cases} \dfrac{V - \min(R, G, B)}{V}, & if\ V \neq 0 \\ 0, & otherwise \end{cases} \\ H &= \begin{cases} 60 \cdot \dfrac{G - B}{V - \min(R, G, B)}, & if\ V = R \\ 120 + 60 \cdot \dfrac{B - R}{V - \min(R, G, B)}, & if\ V = G \\ 240 + 60 \cdot \dfrac{R - G}{V - \min(R, G, B)}, & if\ V = B \end{cases} \end{aligned} \tag{2}$$

2. Applying the thresholds. After that we use the erode and dilate methods to avoid chromatic noise points.
3. Adjust color thresholds by changes of allocated points chrominance, for tracking in variable lighting conditions.
4. Calculating the position of tracked points [5], by knowing the position and trajectory of previous point. Selection the nearest point with appropriate parameters.
5. The trajectory correction to the circle (for mounted camera) or to the ellipse (for perimeter camera) [6].
6. The center estimation (if it is not allocated).
7. Calculation of speed rotation and other main parameters, by the changes of position of tracked points.
8. Save data to the local database.

The algorithm can also be used for a monochromatic camera. For this it is necessary that the central and rotation points can be excluded from the background. In this case, the definition of the threshold occurs without converting frames to other color formats.

The work of the algorithm depends on the camera capability. According to Nyquist – Shannon sampling theorem camera sampling frequency should be 2 times larger than the frequency of the signal. In our case this means that by using the standard camera with 30 frames per second, we can detect a rotation of 15 rotations per second.

However, it is the maximal rotation speed under ideal condition. In real life condition the actual maximal rotation speed is significantly less, because at high speed the image is blurred and not be able to accurately locate the position of the rotating points. If the actual maximal rotation speed for this system is less than 6 times per second, which is suitable for most wind turbines, since the usual speed (without the storm) does not exceed the threshold.

If the system can not detect the position of a rotation point – it does not calculate a rotational speed and the color thresholding, and waits conditions at which detection is possible.

3 Results and discussion

During the experiment, we check the possibility of assessing main parameters of wind turbine, such as speed, orientation, vibration.

Figure 4 shows changing the wind blade rotation parameters over time: current wind speed and passport starting wind speed. It measured by perimeter cameras from different angles.

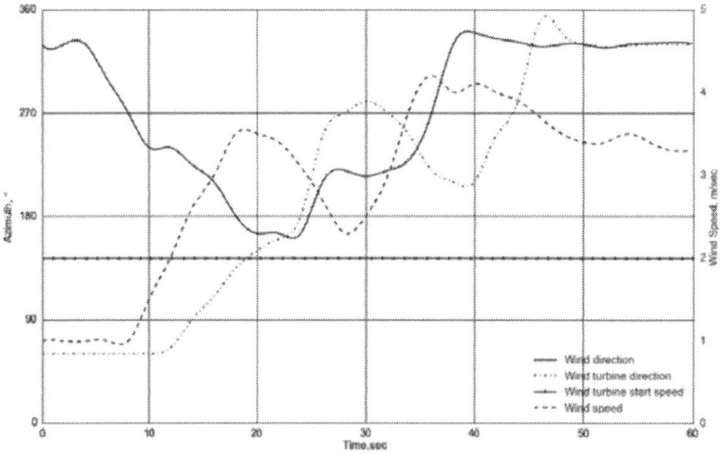

Fig. 4. The registered parameters of wind turbine.

As it can be seen in Figure 4 that mounted camera shows the blade rotation speed more precise, but the perimeter cameras also allow to evaluate the rotation of the wind turbine installation. So, if there is no need for very accurate determination of the blade rotational speed, it is sufficient to use a cameras attached on the perimeter, without making changes in the design of wind turbine.

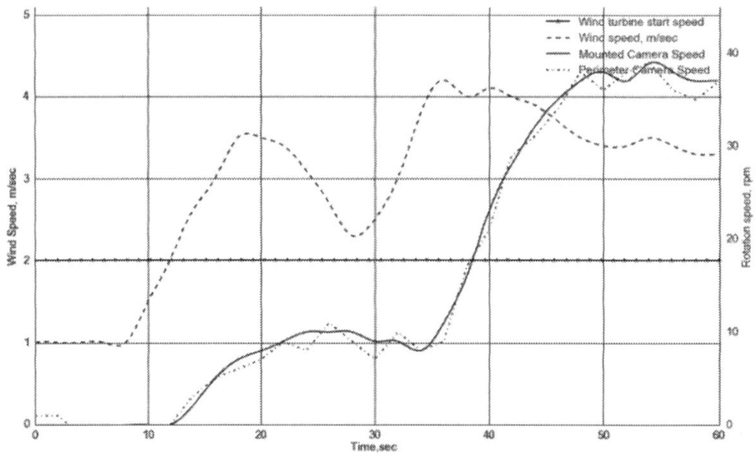

Fig. 5. Position the wind turbine installation, direction and speed of the wind.

To determine these data are sufficient a two standard network video cameras with minimal resolution of 640 x 480 and sampling frequency approximately 30 frames per second. For rapidly rotating generators is necessary to increase the sampling rate of cameras according to the Nyquist – Shannon sampling theorem.

The Figure 5 shows the graphs of the position of the wind turbine blade and wind direction over the time.

The Figure 5 shows that the system can determine the orientation of wind installation in the direction of the wind with sufficient accuracy. It also shows the current and starting wind speed, measured by the weather station. Based on these data it is possible to determine the sensitivity of the wind turbine installation.

4 Conclusion

This article showed that the developed system can detect the rotational speed and position of the wind turbine with sufficient accuracy. The obtained data allow to estimate main parameters of the wind turbine installation and thus allows to callibrate the installation in different conditions without the changes of the design.

The proposed system is a flexible and allows to work with any type of wind turbines with minimal adaptation.

The accuracy of the system depends on parameters of cameras such as image quality and frame rates. Determination of the high-speed rotation with higher accuracy requires more processing power and high speed cameras.

For work at night or in cloudy time, you must provide a lighting installation or, it is possible to use the light or reflective markers instead of the colored marks on the installation.

It is possible to determine variations in the wind turbine, such as vibration and deformation. It will be presented in our further work

References

1. A. Schwarzenegger, *Development of a cost-effective system to monitor wind turbines for bird and bat collisions phase i: sensor system feasibility study* (2007)
2. M. Akhloufi, N. Benmesbah, CRV 2014, Canadian Conference on. IEEE, 246–253 (2014)
3. M. Stokkeland, K. Klausen, T.A. Johansen, ICUAS 2015, International Conference on. IEEE, 998–1000 (2015)
4. S.E. Shcheklein, Yu.E. Nemikhin, S.V. Nevyantsev et al., WIT Transactions on Ecology and The Environment, **190**, 1185–1194 (2014)
5. P. Pérez et al., Computer Vision – ECCV 2002, **2350**, 661–675 (2002)
6. G. Bradski, A. Kaehler, *Learning OpenCV: Computer vision with the OpenCV library* (2008).

E3S Web of Conferences **6**, 03001 (2016) DOI: 10.1051/e3sconf/20160603001
ICSC 2016

Fuel Quality: Challenges to the Sustainable Development of Automobile Transport and Approach to Solution

Elena Magaril[1a], and *Romen* Magaril[2]

[1] Research and Educational Centre ENGEC, Ural Federal University, 620002, Yekaterinburg, Russia
[2] Department of Oil and Gas Processing, Tyumen Industrial University, 625000, Tyumen, Russia

Abstract. The current obstacles to the sustainable development of the vehicle fleet present one of the most serious and complex challenges for technogenic civilization in view of the consumption of rapidly depleting oil and gas resources and the increasing emissions of the products of fuel combustion. The aim of this paper is to study and discuss the impact patterns of the quality of fuel on the environmental and performance characteristics of vehicles and to develop an approach towards improving the quality of fuel as a means of increasing the sustainability of vehicles. The patterns of influence of the chemical composition of the fuel on the temperature conditions in the engine, the emissions of toxic substances and carbon dioxide from the fuel's combustion, the toxicity of the gasoline vapor, carbonization, fuel consumption, engine power and the requirements for a gasoline octane rating have been identified. The use of a fuel additive has been proposed which improves fuel injection in the engines and excludes tarry deposits in the fuel system through the effective detergency and catalyzes the gasification reactions in the engine. The developed fuel additive was shown to have a positive effect by significantly reducing fuel consumption and toxic emissions and improving the performance of vehicles.

1 Introduction

Environmental problems associated with the operation of motor vehicles present one of the greatest challenges of human development at the present time. The high concentration of vehicles in cities directly increases pollution of the air through the toxic substances contained within exhaust gases: carbon monoxide, nitrogen, hydrocarbons, including the most carcinogenic one – benzo(α)pyrene, and soot.

Due to the rapid growth of a world vehicle fleet adapted to liquid petroleum fuels, their consumption is also continuously increasing. The solution to the serious environmental threat posed by traditional fuels and combustion products requires a systematic approach.

One of the elements of a technical culture of operation of the vehicles fuelled by liquid petroleum fuel is the use of additives for various purposes [1–5]. For developing countries that are unable to invest substantially in the refining industry the use of additives for fuels

[a] Corresponding author: magaril67@mail.ru

is the optimal way to rapidly improve the environmental and operational characteristics of vehicles. However, in developed countries, in some cases the use of additives is the only way to maintain optimal engine operation under normal operating conditions.

The proposed study of the impact of fuel quality on the operational and environmental characteristics of vehicles allows the authors to build a concept for the fuel additive which could provide a significant increase in vehicle sustainability.

2 Influence of chemical composition of fuel on the temperature regime in the engine and carbon dioxide emissions

Table 1 presents the lower heating value Q_l for hexane, n-hexene-1, cyclohexane and benzene, according to the reference data [6]. The data show that the heating value per liter of hydrocarbon is the highest for benzene. It should be noted that the calorific capacity per kilogram of aromatic hydrocarbons is lower than that for aliphatic hydrocarbons due to the greater density of the former. Neglecting dissociation of CO_2 and oxidation of N_2, one considers that all the heat released during the adiabatic combustion of hydrocarbon is consumed for heating N_2, contained in the fuel-air mixture, and the reaction products – CO_2 and H_2O. The temperature rise during the combustion of the C_nH_{2n-x} hydrocarbon in a stoichiometric mixture with air can be determined by the equation (1):

$$\Delta T = \frac{Q_l}{\sum m_i c_i}, \tag{1}$$

where m_i – number of moles of the combustion product, c_i – related heat capacity.

Using the reference data [6] on the average heat capacity for CO_2, H_2O and N_2 equal to 56, 45 and 33 J/mol · K, respectively, one can obtain:

$$\sum m_i c_i = 56n + 45(n - 0.5x) + 33 \cdot 3.76(1.5n - 0.25x) = 287n - 53.5x. \tag{2}$$

Then, the temperature rise can be calculated by the equation (3):

$$\Delta T = \frac{Q_l}{(287n - 53.5x) \cdot 10^{-3}} \tag{3}$$

The results of calculations shown in Table 1 reveal that the temperature rise of the products of adiabatic combustion of hydrocarbons at the given number of carbon atoms increases with the degree of hydrocarbon unsaturation and is maximal for benzene. Maximal temperature rise attained during combustion results in increased engine power, however it makes greater demands on the anti-knock characteristic of the gasoline [7].

The greater is the density of the hydrocarbon combusted, the greater is the specific release of carbon dioxide per weight or volume unit of fuel [8–10]. The data presented in Table 1 characterize the specific carbon dioxide emissions per unit of energy produced by the combustion of hydrocarbon, $\frac{G_{CO_2}}{Q_l}$, mg/kJ. Maximal value of $\frac{G_{CO_2}}{Q_l}$ was found for benzene.

Table 1. Lower heating value of adiabatic combustion, maximal temperature rise for the combustion products and specific carbon dioxide emissions for some C_6 hydrocarbons

Hydrocarbon	Lower heating value Q_l, kJ/l	Maximal temperature rise for the combustion products ΔT, K	$\frac{G_{CO_2}}{Q_l}$, mg/kJ
Hexane	30,170	2,150	67.2
n-Hexene-1	30,517	2,215	69.3
Cyclohexane	34,515	2,160	70.8
Benzene	36,390	2,280	82.7

Thus, increasing the content of unsaturated hydrocarbons in the fuel simultaneously with toughening the engine temperature regime increases the specific carbon emissions during engine operation.

3 Toxicity of gasoline

The known Raoult-Dalton relation for ideal mixtures defines the concentration of substances in vapors in equilibrium with the liquid as the following:

$$y = \frac{P}{\pi} x \quad (4)$$

where y – mole concentration of a substance in vapors, x – mole fraction of the substance in liquid, P – saturated vapor pressure of the substance at the given temperature, π – total vapor pressure.

If the concentration of a component with molecular weight M_2 in the gasoline with average molecular weight M_1 is equal to x wt % and its vapor pressure at a temperature T is equal to P kPa, the concentration of saturated vapor of this component of the gasoline can be determined according to the equation (5):

$$C_i = \frac{x/M_2}{x/M_2 + (100-x)/M_1} \cdot \frac{P}{101.3} \cdot \frac{M_2 \cdot 273}{22.4 \cdot T} \cdot 10^6 \ мг/м^3 \quad (5)$$

The vapor toxicity of a given gasoline component can be characterized by C_i/MAC_i ratio, where MAC is the maximal allowable concentration, while the total vapor toxicity of a multi-component gasoline mixture is equal to $\sum C_i/MAC_i$.

Table 2 shows the results of calculation of the toxicity of vapors of n-hexane and its binary mixtures with n-hexene-1, cyclohexane and benzene, made on the base of the average daily *MAC* values stated in Russia.

Table 2. Vapor toxicity of n-hexane* and its mixtures with the second component (at 101.3 kPa, 30°C)

Component	Vapor pressure, kPa	Component concentration in vapors mg/m³	Vapor toxicity in *MAC* units	Vapor toxicity relative to hexane
–	25.43	868,375	57,890	1.00
component concentration in a mixture with hexane of 1 wt %				
n-Hexene-1	30.96	10,565	181,5930	3.14
Cyclohexane	16.20	5,530	61,250	1.06
Benzene	15.91	5,420	111,485	1.93
component concentration in a mixture with hexane of 5 wt %				
n-Hexene-1	30.96	52,800	676,070	11.68
Cyclohexane	16.20	27,615	74,660	1.29

Benzene	15.91	25,990	314,740	5.44

*average daily *MAC* for the atmospheric air at the settlements for n-hexane is taken to be equal to 15 mg/m^3

The data show that at the same content in the mixture n-hexene-1 and benzene heighten the toxicity of the vapors to the greatest extent. Note, however, that benzene is very stable under normal conditions and practically does not enter into chemical reactions, while n-hexene-1 is sufficiently rapidly oxidized with oxygen in the air and removed from the atmosphere. In view of this, benzene is the least desirable component of the gasoline and limiting its content in a fuel set to no more than 1 vol. % by current European standards EN 228:2012 as well as by the Worldwide Fuel Charter [11] is quite justified. It should be noted that motor transport is a source of 80–90% of benzene in the atmosphere [12–14]. Oxidation products of benzene in the body, especially phenols and polyphenols, have a toxic effect and disrupt enzymatic processes in the cells and cause severe poisoning of the body or even death. Oxidation of benzene homologs leads to the formation of less toxic compounds.

To reduce losses from evaporation of gasoline and its negative environmental and economic consequences the saturated vapor pressure of gasoline should be limited. Essentially, with no change in chemical composition, the saturated vapor pressure of the gasoline can be reduced by introducing into it a surfactant additive which reduces the percentage of gasoline in the surface layer [3, 4].

4 Carbonization and its impact on environmental and performance characteristics

During the process of vehicle operation carbon is deposited on the cylinder walls and the pistons. Aromatic hydrocarbons possess the greatest propensity for forming carbon deposits. The authors investigated the dependence of pyrocarbon formation in the pyrolysis of individual hydrocarbons in an inert carrier on the chemical structure of the hydrocarbons. At a temperature of 1123 K the rate of the pyrolytic carbon formation from benzene is 7 times higher than that from cyclohexane.

Condensed (polycyclic) aromatic hydrocarbons are the precursors of carbon deposit during its formation. Benzene, as well as other aromatic hydrocarbons, has the highest propensity to form benzo(α)pyrene compared to hydrocarbons of other classes [15–17].

Table 3 illustrates the impact of hydrocarbons on the content of benzo(α)pyrene in exhaust gases. It is seen from the data presented that at the same number of carbon atoms alkanes and cycloalkanes give amount of benzo(α)pyrene which is, by the order of magnitude, smaller than that given by benzene.

Table 3. Concentration of benzo(α)pyrene in exhaust gases during combustion of hydrocarbons (excess air coefficient is 1.15)

Hydrocarbon	Concentration of benzo(α)pyrene in exhaust gases, ppm
Hexane	1.0
n-Hexene-1	1.0
Cyclohexane	1.1
Benzene	28.6

At peak temperature concentrations of carbon and nitrogen oxides formed during the fuel combustion in the engine are maximal and futher decreasing the temperature "freezes" the concentrations of the gases formed [4]. Thus, according to estimates for the formation of nitric oxide from nitrogen and oxygen within the air, when decreasing the temperature from 2500 to 2000 K the reaction rate decreases by a factor of $2.2 \cdot 10^6$ and the equilibrium concentration of nitrogen oxide reduces by a factor of 7.1. When the temperature drops from 3000 to 2000 K, the equilibrium concentration of carbon monoxide at a pressure of 1 MPa in the products resulting from the combustion of hydrocarbons in oxygen reduces from 23 to 18 vol. %.

Since an increase in the content of unsaturated and aromatic hydrocarbons in the fuel toughens the temperature conditions in the engine due to both the higher temperature developed during combustion and increased carbon deposits formation, which hinders the process of heat-removal through the wall, the greater the degree of unsaturation of the fuel is, the greater are the emissions of oxides of carbon and nitrogen during the vehicle operation.

As the carbon deposits form, the friction force between the piston and the cylinder sharply increases and this increases fuel consumption. Another reason for increasing fuel consumption is the emergence of so-called "surface ignition" – preignition which leads to an increase in pressure of the gaseous mixture in the compression stroke as a result of the reaction heat release. This increases the energy consumption for compression and results in loss of engine power. In addition, the low thermal conductivity of carbon deposit leads to an increased temperature of fuel combustion [3, 4, 7, 18, 19], which increases the requirements for the anti-knock resistance of the gasoline. Deterioration of heat removal from the combustion chamber due to increased carbonization results in an increase in temperature when exhaust gas is released, which may lead to burnout of exhaust valves.

5 Soot emissions

The micro-diffusion combustion of fuel in diesel engines results in the formation of soot, the emission of which is a significant environmental problem.

Figure 1 shows a model of combustion of droplets of diesel fuel in an engine. Fuel vapors diffuse to the flame front continuously heating. A deep pyrolysis of fuel molecules occurs and soot particles form. Some of the soot particles passing through the front flame burn out. However, the passing time for the largest particles is not enough to burn and they are emitted with the exhaust gases. The soot output increases as the C:H ratio in the fuel increases [20–22] and so too does the fuel density and the content of heavy fractions. The highest soot output is due to thermal decomposition of bi- and tricyclic aromatic hydrocarbons; monocyclic aromatic hydrocarbons give lower output and alkanes and cycloalkanes – even less.

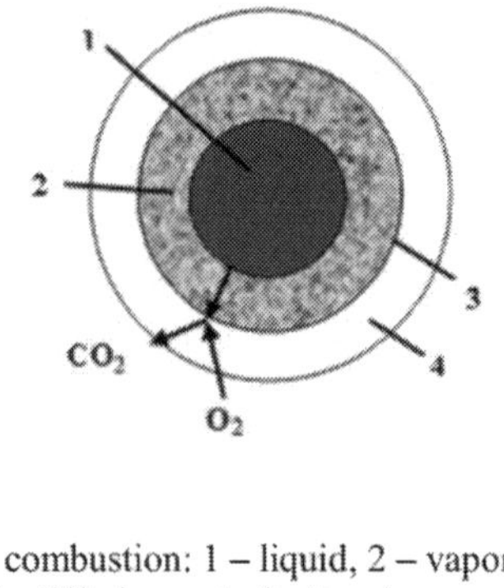

Fig. 1. Scheme of diesel droplet's combustion: 1 – liquid, 2 – vapor layer, 3 – combustion zone, 4 – layer of oxygen and carbon dioxide diffusion and afterburning.

Since soot formation increases with an increase in aromaticity and the molecular weight of the aromatic hydrocarbons, regulation of the polycyclic aromatic hydrocarbons' content in the diesel fuel is necessary and justified. The following ways to reduce soot emissions from diesel engines are possible:

1. Decreasing the formation of soot particles. This can be achieved by increasing the total surface of the fuel droplets. As a result of this the rate of evaporation of the fuel increases and droplets' residence time in the preflame zone reduces. The work of spraying consists of the work done during the injection to overcome aerodynamic resistance and the work to create a new surface equal σS, where σ – the surface tension on the fuel/air boundary, S – total surface of the droplets. At the constant aerodynamic resistance a reduction in surface tension leads to an increase in the total surface of the fuel droplets.
2. Increasing the degree of combustion of soot particles. This is possible if the soot particles will comprise a carbon gasification catalyst.

6 Increasing sustainability of automobile transport by means of a multifunctional additive for fuels

6.1 The concept of the additive providing improved sustainability of vehicles

To reduce carbonization in an engine with the appropriate temperature mitigation conditions in the engine, to decrease toxicity of exhaust emissions and fuel consumption it is necessary to improve fuel spraying, which can be achieved by introducing a surfactant that lowers the surface tension of the fuel droplets at the interface with air. Preventing the non-evaporable fuel droplets from being reached the wall will minimize carbonization. Improved spraying will reduce soot formation in diesel engines as well.

The introduction into the engine, along with the fuel, an element which could provide catalysis for gasification reactions, will not only contribute to the elimination of carbon deposition and soot burnout, but also greatly reduce emissions of products of incomplete oxidation and condensation of hydrocarbons of the fuel, including polycyclic aromatic hydrocarbons.

For regular engine supply with the gasoline, it has to possess high detergency to prevent occurrence of tarry substances in the gasoline pipeline. It can be achieved by introducing a surfactant. Reduction of losses of the gasoline from evaporation may also be achieved due to the surface-active effect of the additive introduced.

Thus, to achieve the desired integrated effect of improvement to both the environmental and operational characteristics of vehicles the additive should combine the properties of the surfactant and the gasification catalyst.

6.2 Results of the influence of the versatile additive on the properties of fuels

The authors have developed a versatile additive for gasoline and diesel fuels with a high surface activity, containing within the structure an effective catalyst for gasification reactions. The additive can be applied to the fuel in ultra-small concentrations: 9.25 mg per 1 kg of gasoline and 27.75 mg per 1 kg of diesel fuel.

The developed additive significantly increases the detergency of fuels. The results of the additive's influence on detergency are shown in Figure 2. Introduction of the additive into gasolines reduces the saturated vapor pressure by 15–20% which leads to a decrease in gasoline loss due to evaporation of 17–30%.

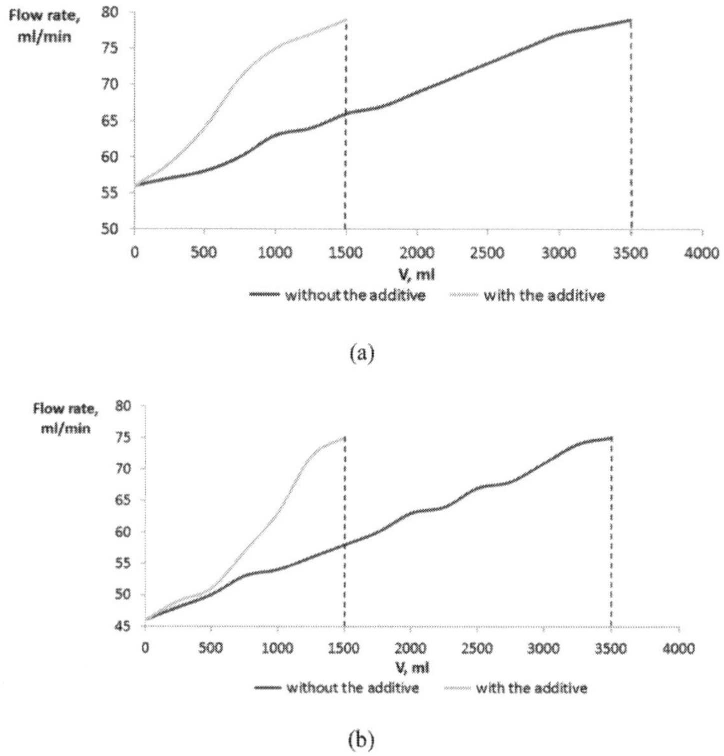

Fig. 2. Influence of the additive on the detergency of fuels: (a) gasoline (concentration of the additive is 9.25 mg/kg); (b) diesel fuel (concentration of the additive is 27.75 mg/kg).

The additive significantly reduces carbonization due to improved fuel spraying and catalyzes the gasification reactions of polycyclic aromatic hydrocarbons (PAH), including benzo(α)pyrene, and carbon deposit. As a result, PAH emissions are reduced by up to 95% and carbon deposits are removed almost completely.

Enhancing heat-removal through the wall by eliminating carbon deposit facilitates the combustion process in the engine and reduces emissions of gaseous toxic substances: nitrogen oxide emissions reduce by 20.6% in a diesel engine and by 26.5% in a gasoline engine, CO emissions drop to 14.6% (depending on the excess air coefficient). Increased completeness of combustion of hydrocarbon fuels results in emissions reducing by 35.5–37.2%. Eliminating carbonization reduces the non-productive power expenditure and leads to a reduction in gasoline consumption by 10–12% (Figure 3), diesel fuel – by 5–7%. As a result, a reduction in the specific fuel consumption leads to a corresponding decrease in carbon dioxide emissions. Elimination of carbon deposit formation in the engine reduces the requirements for gasoline octane number by 7–12 points.

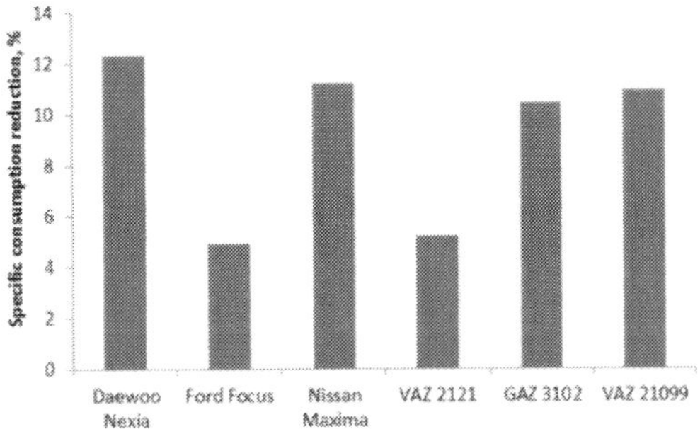

Figure 3. Influence of the additive on the gasoline specific consumption.

Due to the surface-active influence of the additive, which improves the fuel spray process, combined with its catalytic effect, soot emissions are reduced significantly. In the diesel engine Zil-64 bench test a reduction of soot in the exhaust gases of the engine at maximum load was 50% (from 0.36 to 0.18 mg/l). Thus, the developed versatile additive exhibits effectiveness for both gasoline and diesel fuels, improving the environmental and operational characteristics of fuels concurrently.

7 Conclusion

Patterns of influence of the chemical composition of fuels on the environmental and operational characteristics of vehicles were identified. An increase in the content of aromatic hydrocarbons was shown to lead to a toughening of the temperature regime in the engine due to both a rise in the maximum temperature of fuel combustion and to increased carbonization. As a result, the specific consumption of fuel, emissions of toxic substances, soot and greenhouse gases increase. Moreover, this also increases the requirements for the octane number of gasoline consumed.

Therefore, measures are needed to reduce the content of aromatic hydrocarbons in fuels which requires the development of hydrocracking and hydrorefining capacities in refining industry.

A concept was formulated for a fuel additive which provides an opportunity to increase sustainability of automobile transport due to the integrated positive influence on the fuel's properties. The additive, combining the properties of a surfactant and a catalyst for gasification reactions, greatly improves the environmental properties of fuels, reducing their specific consumption, gasoline loss from evaporation and reduces the need for high-octane gasolines. Use of the additive can significantly mitigate the negative impact of the high content of aromatic hydrocarbons in fuels which is a serious issue for a number of developing countries.

Acknowledgement

This research was supported by Act 211 Government of the Russian Federation, contract № 02.A03.21.0006.

References

1. J. Bennett, in: *Alternative fuels and advanced vehicle technologies for improved environmental performance. Towards zero carbon transportation,* 165–194 (2014)
2. A.M. Danilov, Chem. Tech. Fuels Oil+, **47 (6)**, 470–484 (2012)

3. E. Magaril, WIT Trans. Ecol. Envir, **147**, 373–384 (2011)
4. E. Magaril, International Journal of Sustainable Development and Planning, **8** (**2**), 231–245 (2013)
5. S.P. Srivastava, J. Hancsók, *Fuels and fuel-additives* (New Jersey, 2014)
6. Yu.M. Zhorov, *Thermodynamics of chemical processes: petrochemical synthesis, processing of petroleum, coal, and natural gas* (Moscow, 1987)
7. E.R. Magaril, R.Z. Magaril, V.G. Bamburov, Combust. Explo. Shock+, **50** (**1**), 75–79 (2014)
8. E. Magaril, WIT Trans. Ecol. Envir, **190** (**2**), 821–832 (2014)
9. E. Magaril, Environ. Sci. Pollut. R., **23** (7), 6793–6802 (2016)
10. E. Magaril, International Journal of Sustainable Development and Planning, **10** (**6**), 880–893 (2015)
11. *Worldwide Fuel Charter*, 5th edn, (Belgium, 2013)
12. C. Guerreiro, F. De Leeuw, V. Foltescu, J. Horálek, *Air quality in Europe. EEA (European Environmental Agency) Report* (Luxembourg, 2014)
13. O. Harrop, *Air quality assessment and management: a practical guide* (London, 2002)
14. L. Jinawa, S. Thepanondh, International Journal of GEOMATE, **11** (**2**), 2341–2347 (2016)
15. A. Candeli, G. Morozzi, M.A. Shapiro, in: *Mobile source emissions including policyclic organic species*, 29–47 (Netherlands, 1983)
16. A. Comandini, T. Malewicki, K. Brezinsky, J. Phys. Chem. A, **116** (**10**), 2409–2434 (2012)
17. K. Ravindra, R. Sokhi, R. Van Grieken, Atmos. Environ., **42** (**13**), 2895–2921 (2008)
18. E. Magaril, WIT Trans. Built Env., **130**, 437–485 (2013)
19. E. Magaril, International Journal of Sustainable Development and Planning, **10** (**1**), 100–108 (2015)
20. L.M. Li, P.B. Sunderland, Combust. Sci. Technol., **184** (**6**), 829–841 (2012)
21. P.A. Tesner, S.V. Shurupov, Combust Sci Technol., **105** (**1–3**), 147–161 (1995)
22. *Chemistry and physics of carbon* (New York, 1965)

E3S Web of Conferences 6, 03002 (2016) DOI: 10.1051/e3sconf/20160603002
ICSC 2016

Energy service agreement as a tool of the program implementation aimed at raising of energy efficiency: challenges and opportunities

Anastasia Tupikina[1a], and *Sergey* Chernov[2]

[1]Teaching assistant of the Department of the production management and energy economy, Novosibirsk State Technical University, Russia
[2]Candidate of Economics, Associated Professor, Head of Department of the production management and energy economy, Novosibirsk State Technical University, Russia

Abstract. In this article relevance of energy service agreement as a tool of energy efficiency raising has been proved. On the basis of analysis of researches aimed at energy service agreements, legislative base and procedural framework, examples of implementation of energy service agreements the key challenges have been defined, slowing down the development of energy services market in Russia. Possible ways of solving these problems have been shown and the necessity of complex approach to dealing with these issues has been drawn.

Today relevance of the issue of energy efficiency improvement in the Russian economy is practically assured. Researches devoted to energy efficiency in Russia proved that energy output ratio of the Russian economy is higher than in the American one by 2–3 times and by 3,5–6 times higher than in West European and there is a stable trend in raising of energy output ratio [1].

Issues of energy saving and raising of energy output ratio got their urgency after the first Energy crisis in 1970. The richest experience in these issues belongs to such countries as Japan, the USA and the countries of the Eastern Europe. The usage of the energy saving agreement as a tool of energy efficiency raising is the standard of practice abroad.

In Russia the interest to the problem of energy efficiency raising has appeared not so long ago: the first legislation in the sphere of energy saving appeared at the end of 1990, but energy service hasn't been widely spread, in spite of apparent advantages of this type of interaction:

– Feasibility of using the competencies of experts in the sphere of energy efficiency raising.
– Advanced abilities of fund raising, connecting with the fact that in the process of implementation of performance agreement this issue is dealt by energy saving company as a rule.

Experts in the sphere of energy saving bind the slow growth of energy service market in Russia with the problems appeared in the process of implementation of energy service agreement. For better understanding of future development of energy service market it's

[a] Corresponding author: tupikina.aa@mail.ru

necessary to consider the given below problems, find out the causes of these problems, figure out the possible ways of their solving.

1. Lack of state support

The analysis of the foreign experience in the sphere of energy efficiency raising proved that in the most developed countries initiating stakeholder is the state. These countries are Japan, the USA, Germany, Austria, Sweden and the other countries of Europe.

Apparently, that in the issue of energy service market development as a tool of energy efficiency raising the state participation is necessary. State support must be expressed in the form of information and legislative activity.

In the part of information activity, the experts from their personal experience noted, that in spite of state caring out large and expensive information campaign, aimed at explanation of statements of Federal Law 261, from 7 of 10 cases, it's necessary to explain to managers of different levels the aim and mechanism of energy service implementation from the very beginning [2], that characterizes the efficiency of information support.

As for the Russian legislation in the sphere of energy service the majority of researchers noted down, that its imperfection and the lack of clarity in many issues, in legislative and tax support of energy service agreements.

Apart from legislative and information support, active state participation in the development of energy service market by introducing the tax remissions, acting as a guarantee before financial institutions, direct investment with the help of mechanism of state-private partnership would allow to sort out the other problems, listed below.

2. Difficulties in fund raising

Market research of energy service in Russia proved, that the majority of implementing now projects in energy efficiency improvement are cost efficient, because the projects with short payback period are of priority.

Nevertheless, considering the process of energy efficiency improvement as transition to the new technological level, it's necessary to emphasize the projects provide modernization of equipment and technologies, it implies heavy expenses. This statement is particularly true for enterprises, where the large losses of power resources are connected with deterioration of equipment.

Connecting the abovementioned facts, nowadays in Russia the program aimed at fund raising of activities on energy efficiency improvement in the frame of energy service agreements is urgent.

The key problem is that the major part of energy service companies are average and small companies which do not have sufficient their own capital base, that's the reason why this type of financing is not used. It ought to be remarked that while attracting the leveraged financing some difficulties appear such as considering this type of business as very risky by financial institutions, that's why interest rates for such loans are high enough. Moreover, the experts consider the projects with interest rate more than 14% are money loosing [3, 4].

Coming back to the issue of lack of state support, it should be noted that state participation as one of the parties of energy service agreement, would allow to solve the problem of financing energy service agreements to some extent.

3. There are large risks for energy service companies not to get payment

Energy service agreement for energy service companies is defined like investment, consequently, besides some specific risks energy service companies take some conventional risks, which are attributed to investment projects.

High risks with the given type of activity is one of the reasons for some difficulties' occurrence in fund raising, which have been mentioned above. Financial companies are not ready to accept the risks, connecting with not payback back the investment money in a full volume.

Insurance mechanism of energy service companies has been used as a tool of reduction the risks, appearing while implementation of measures in energy efficiency raising for a long time. Moreover, issue of insurance is one of the key points in foreign energy performance agreements, but while adapting energy service agreements to the conditions of Russia, this item was excluded from it. In such a way, the risks exist in energy saving projects abroad increase in many times because of non binding nature of contract relations and not so great experience of native energy service companies in this sphere [5].

4. Impossibility of defining the proper value of energy saving due to the difficulties appear with validity of datum line

Urgency of the issue of defining the base line of energy consumption is determined by the fact that the base line is the base of all the calculations made by the agreement, consequently the given problem enlarges the risk of non payment money back [2].

Within the framework of the researches, the analysis of performed in Russia projects has been made, which revealed some peculiarities of defining the base line of energy consumption [6]. They are:

1. Sensitivity analysis of projects showed that the slightest deviation of the value of base line can lead to shortage in economy up to the occurrence of situation when a project becomes money loosing. The example of such analysis is given in Table 1. The project of modernization of heating net-work of one heat power plant in Western Siberia has been chosen as an object.

Table 1. Sensitivity analysis of project by changing the effect (value of economy) caused by corresponding changing of base line.

Changing of base line, %	-10,7	-7,14	-3,57	0	+3,57	+7,14	+10,7
Changing of effect, %	-15	-10	-5	0	+5	+10	+15
NPV, th, RUR	-5 165	-1 539	2 077	5 575	9 072	12 596	16 06

2. Comparison of variables of project efficiency depending on the method of defining of the base line proved that the choice of methodology of its defining impacts greatly on the financial results of the project. The analysis of variables' comparison of the same project is given in Table 2.

Table 2. Comparison of variables of project efficiency depending on the method of defining the base line.

Variable	Method 1	Method 2
Net present value (NPV), mln. RUR	100,1	5,575
Internal rate return (IRR), %	51	17
Payback period , years	3,11	5, 21
Discounted payback period, years	3,62	7,24

Problem solving of defining of the base line can be development of universal methodology suitable for any type of objects, by the way considering branch and regional particularity of projects. The most prospective and full one at the present moment is the Standard «Measurement and verification of energy efficiency» issued by the Russian association of energy service companies [7]. The given standard was called by experts «revolution of the branch» [8] that gives hope that application of the standard will allow experts to solve some problems, appearing in the process of energy service agreement implementation.

5. Distribution of money value of extra economy of energy resources

Existing practice of energy service agreement implementation in Russia showed that one of the most urgent problems related to interaction between client and energy service company is the problem of distributing of money value of economy of fuel energy resources (FER), which appears as a result of measures implementation aimed at energy efficiency raising.

The way of developing of contract price should be explained. While making an agreement two variables are given by energy service company: variable of energy resources saving (given in percentage of current consumption) and proportion of money economy, which must be paid to energy service company during the whole period of project implementation. Payment given to energy service company can be expressed by the formula (1):

$$P^{ESC} = \alpha^{ESC} * \Delta V^{FER} * T, \quad (1)$$

where P^{ESC} – payment, got by energy service company,
α^{ESC} – proportion of money economy, announced by energy service company,
ΔV^{FER} – saving of fuel energy resources FER in physical terms (as difference between basic and factual levels of consumption);
T – price per unit of FER (tariff).

In the process of energy service project implementation three variants of economy distribution can appear:

1. Energy service company does not reach the declared level of economy – energy service company doesn't get payment and pays the forfeit according to the terms of agreement.
2. Achieved level of economy corresponds to declared one – energy service company gets payment according to the terms of agreement.
3. Achieved level of economy exceeds the declared one.

Apparently, while distributing amount of extra economy, conflict of interests of client and energy service company appears:

– From the point of view of the client, energy service company declares necessary for it amount of payment, consequently the money value of economy, which exceeds the level declared by energy service company, must belong to the client.

– The opposite interest appears from energy service company: the client has only declared in the agreement level of economy, consequently, everything, which has been got extra – it's payment for energy service company.

In such a way, the distribution of amount of extra economy must be practically done by compromise. Apparently, the conditions of distribution of extra economy must be written in agreement in order to avoid conflicts between parties.

Nowadays no legislative and methodical guidelines exist, connecting with distribution of extra value of economy.

RF Government regulation No. 636 contains the condition about the percentage of extra economy, that this percent paid to energy service company can't exceed fixed percent of economy in the terms of money of corresponding expenses of client for getting energy resource, stipulated in the agreement [9]. By the way, the given regulation concerns only state and municipal institutions, moreover, the given condition is considered to be limiting but not sufficient for determination of conditions of distribution of extra economy, that's one more time underlines imperfection of legislative framework of energy service agreements.

Analyzing the mentioned above problems, it should be noted, that, they are interconnected with each other, and it's not excluded that, joint solving some of them will have synergy effect.

For example, the problem of fund raising is closely connected with large risks of energy service projects. Meanwhile, involving the state support will help to solve these problems as a complex. The problem of imperfection of legislative framework and normative base causes the described difficulties in defining base line and distributing of extra economy, which in its turn, enlarge the risks of energy service companies within the projects.

In such a way, it's apparent, that for development of the market of energy service in Russia, complex solution of the problems is necessary. Dealing with the problems the following directions can be figured out:
1. Improving of normative base connecting with concluding agreements with energy service companies.
2. Development and integration of methodical statements by the most difficult agreement processes: defining of base line, validity of measurements, distribution of extra economy and etc.
3. Stimulating of interest of financial and insurance organizations to implement new loan and insurance products.

The article has been prepared within the frame of the subject plan of Novosibirsk State Technical University TP-PM&EE-2_15.

References

1. I.A. Bashmakov, *Regional policy of energy efficiency raising: from problems to their solving* (1996)
2. S.V. Tsakunov, Energy saving, **3** (2012)
3. F.F. Lavrinenko, *Materials of advance training course «Energy service agreements: preparation and management», Moscow, 10.04–20.04.2013.*
4. G.N. Ivanov, Energosovet, **2** (**15**) (2011)
5. O.V. Uliankin, Achievements of university science, **4**, 169–174 (2013)
6. A.A. Tupikina, S.S. Chernov, Investments, **10**, 106–112 (2015)
7. Measurements and validity of energy efficiency: standard, Russian association of energy service companies. – Stated by Council of Association of energy service companies «RAESC» 16 September 2014, (2014)
8. EnergoSovet.Ru – energy saving website (2014), http://www.energosovet.ru/news.php?zag=1413965575
9. RF State Regulation, dated by 18 August, 2010 No. 636 «About requirements for conditions of energy service agreement and about peculiarities of defining primary (maximum) contract price of energy service», http://www.rg.ru/2010/12/14/energoservis-site-dok.html

E3S Web of Conferences 6, 03003 (2016) DOI: 10.1051/e3sconf/20160603003
ICSC 2016

Energy-efficient architecture of industrial facilities associated with the desalination of sea water

Timur Gazizov[1a], and *Irina* Melnikova[1]

[1]Moscow State University of Civil Engineering, 129337, Moscow, Russia

Abstract. The article offers an actual solution of a problem of drinking water shortage in the territory of the Crimean coast, in the city of Sudak, Autonomous Republic of Crimea, Russia. The project includes a development of energy-efficient architecture, its implementation in industrial facilities, such as stations for seawater desalination and an active use of alternative energy sources.

1 Introduction

Today there are industrial enterprises that are directly related to the functioning of cities and to existence of human society in them. For example, a modern city cannot exist without incineration plants, thermal power stations, treatment plants, etc. Such enterprises include plant for desalination of sea water. Especially necessity of uninterrupted operation of such enterprises increases in the case of industrial and natural disasters. Suffice it to recall the tragedy in Japan, 2011, when the regular major earthquake unbalanced a large part of the Japanese residential areas. What is particularly important – in the first days after the tragedy all bottled fresh water was sold out, even in areas that located far from the most affected places. Value of drinking water increases by tens and hundreds of times under such conditions. Today, fresh water in industrialized countries is worth more than oil even against the background of a peaceful and prosperous life in cities. More and more scientists tend to believe that no oil and gas, but clean fresh water will become a major problem of the XXI century. In arid regions of the world water was always number one and was the cause of many wars and armed conflicts in the territories, which was poor of natural fresh water resources. A good example of such conflict is the "War for water 1964". Thus, engineering of such objects, in particular enterprises for the production of drinking water has a great importance in today's dynamically developing world.

In Russia, in urgent need of large amounts of fresh water is the Caspian Sea and the Crimea, so the design of industrial plants for desalination of sea water in these areas is an extremely important proposal that can improve a functioning of residential areas.

The importance of such projects for the Crimea is completed by a need of full functioning of its resort component. In particular, in Sudak, which is one of the major resort cities of Crimea, fresh water is supplied for a few hours a day.

[a] Corresponding author: timur.gazizow@mail.ru

2 Station Project on seawater desalination

The authors offer the Station Project on seawater desalination for the city of Sudak. A district chosen for the station building is optimal for communication with the city because it is located very close to the city and at the same time it is separated from residential areas by a large coastal rock. This arrangement of the industrial object associated with the city's fresh water providing extremely beneficial strategic reasons. At the same time the station has a direct link with runs along the coast highway. The lay of the land allows supply water to the city with minimal use of pumping equipment (Figure 1).

Fig. 1. Situational plan.

Modern methods, technologies and the general trend of development of the technologies for desalination were studied and the most promising of them were identified, namely: the reverse osmosis technology.

By comparison with the long-used "thermal distillation" technology (multi-residue), used since the beginning of the last century, and on which the main emphasis in comparison technology has been made, the technology of "reverse osmosis" (passing sea water through a semipermeable membrane against high pressure) compares favourably with its constructive simplicity, relative cheapness of production and operation.

Reverse osmosis plants is actively used in the world practice the desalination of sea water from the end of the last century. The indisputable advantage of the reverse osmosis method in comparison with its thermal distillation is its small-size, which allows the use of reverse osmosis in the premises of any configuration (Figure 2),

Fig. 2. Examples of reverse osmosis plants of desalination of sea water.

without using of a huge number of hydraulic components, communications, boilers and tanks and other containers, which in the "old technology" was in abundance (Figure 3).

Fig. 3. Examples distillation stations, seawater desalination.

The project involves the design of the station with a complete set of technological operations, shops and facilities needed for a full cycle of desalination of sea water from a water extraction of from the Black Sea its desilting, pre-treatment, ultrafiltration on hollow fibers and disinfection, membrane filtering through the installation of "reverse osmosis", further mineralization and following the fresh water discharge into the existing system of water supply system of Sudak. Perhaps the accumulation of fresh water in the containers located on a coastal area, in the vicinity of the station, and further transportation of fresh water main pipeline to the areas with poor own water resources. For example, an end of a holiday season. The Project provides solution to the problem of waste removal "brine" from the station, namely its multi-stage drying and further use of the resulting dry mixture in various industries, such as fertilizer. The project also provides for the availability an alternate supply, capable of providing a work station in emergency situations without a prolonged power (Figure 4) [1].

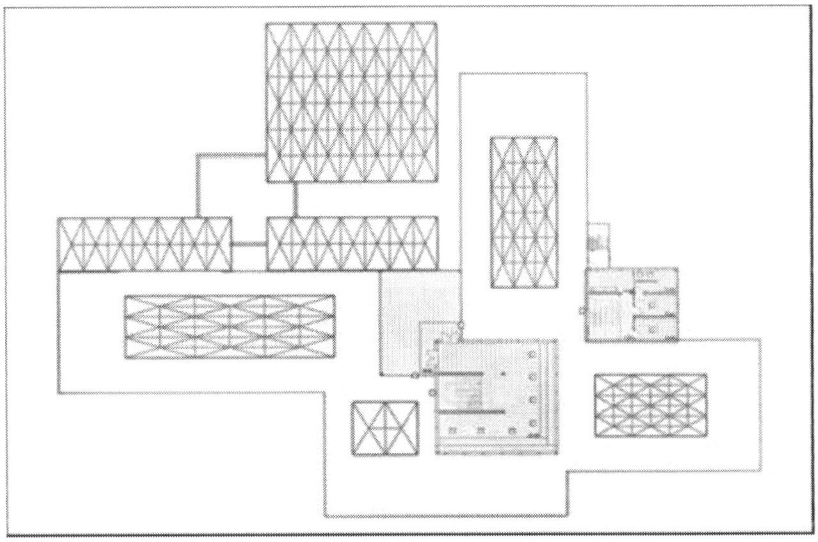

Fig. 4. Plan of the 3rd floor of the station to the placement of solar modules.

Choice planning solutions station based on the specific process for desalination of sea water by the method "Reverse Osmosis" (the transmission of sea water through a semipermeable membrane under high pressure), and the possibility of placing alternative sources of supply in the body of this industrial facility, namely, distribution of polycrystalline solar modules of the third generation (Figure 5).

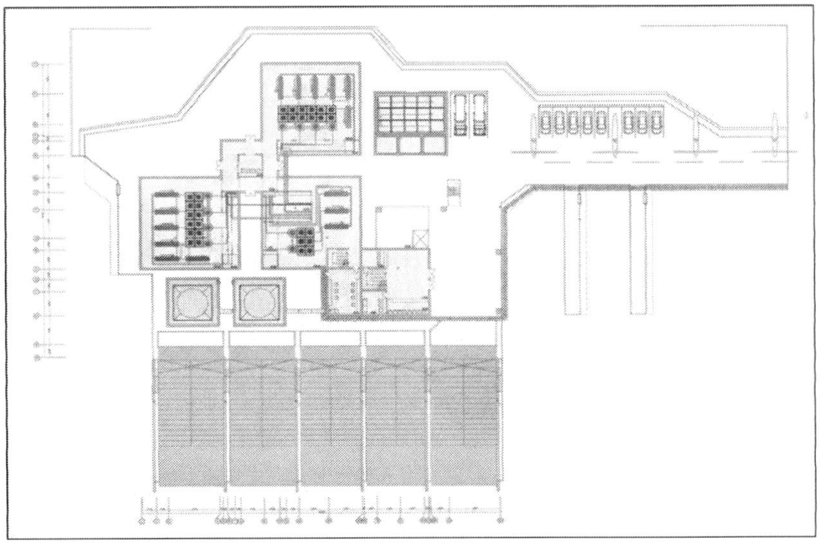

Fig. 5. The plan of the 1st floor of the station.

Climatic conditions of the Crimea make it possible to use solar panels is very effective even without any preliminary calculation. Shop organization for manufacture of the stations complemented by the development of administration and accommodation unit, combining the current level of control station and the comfortable stay for the working personnel. This unit has a convenient functional connection with the shops, and at the same time it is located on the side entrance to the station. The need of walking any visitors who came to the station administration, excluded from its territory. The station building was designed taking into account the partial bearing on the seabed and the coast rocky slope of the shoreline (Figure 6).

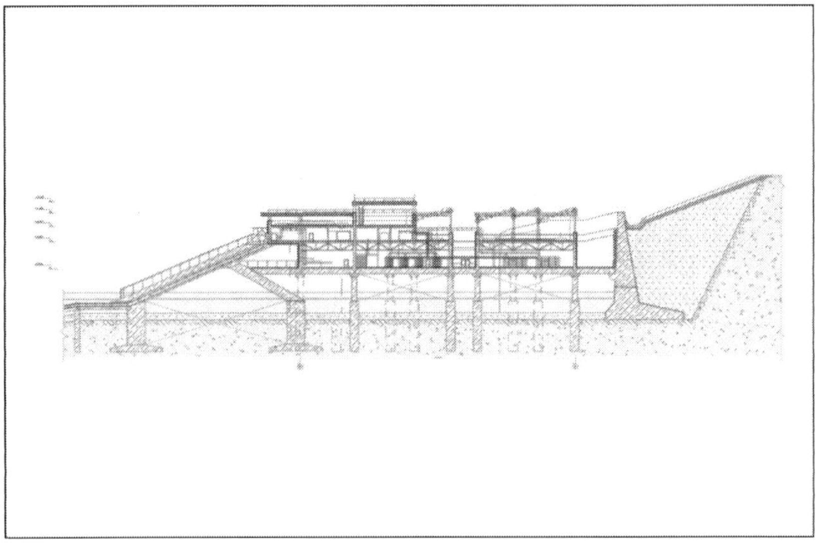

Fig. 6. Constructive section of the station.

The structural building system involves the use of reinforced concrete structures bearing elements of walls and ceilings, using metal trusses at overlapping large spans the shop of the station, installation of foundations using "compressed-air sinking" technology, availability turns off the links and other measures when designing in the area having a high seismic activity. There was also offered the option of using non-metallic reinforcement in concrete structures bearing plant [2]. The roof design involves the installation of rotary polycrystalline silicon third-generation modules with an efficiency of 14–16%, with a total area 920 sq.m., allowing for maximum use of solar radiation energy of about 1350–1500 Vat/sq.m., coming to the surface of the station roof. Such a choice of solar modules provides to the station to 14% efficient power of her work – "in a conservation of energy mode." Also at the territory object have been installed wind turbines that allow to a total of about 8% efficient power needed for desalination plants [3]. According to our calculations, in the absence of electricity, residents and vacationers of the city of Sudak can be provided with fresh water in a volume of 10 liters/day per person, and it is during a station work alone alternative sources of energy in the height of the season.

3 Conclusion

The use of energy-efficient systems in the structure of industrial projects in these conditions is not just effective measure, but it is extremely necessary because they provide significant energy-intensive enterprises with cheap renewable energy resources. As already mentioned, such a system at the industrial enterprises of strategic importance, in particular, associated with fresh water obtaining, can be indispensable in emergency situations, in the event of terrorist attacks and natural disasters (Figure 7), (Figure 8). This variant can be an universal solution in the design of such construction in places where building directly on the coastal territory of the land is impossible.

Fig. 7. General view from the north-west side of the station.

Fig. 8. General view from the south-east side of the station.

References

1. V.N. Slesarenko, *Desalination of sea water* (1991)
2. S.V. Dyatkov, A.P. Mikheyev, *The Architecture of Industrial Buildings: the textbook for high schools* (2008)
3. V. Germanovich, A. Turilin, *Alternate sources of energy, Practical design for the use of wind, solar* (2011)

Problems of energy-efficient construction in Russia

Liubov Shilova[1], *Leonid* Shilov[1a], *Andrey* Atamanov[1], and *Ilya* Ermakov[1]

[1]National Research Moscow State University of Civil Engineering, 129337, Moscow, Russia

Abstract. The article considers the world practice in the field of energy-saving construction technologies as well as the present-day situation in Russia. Also the authors consider problems associated with the growth of energy consumption and electricity generation, raise questions concerning the environmental problems of energy production. The recommendations on the reduction of energy consumption are given.

The electricity production is a complicated labour-consuming process depending on many factors from world economic market situations to the power resources limitedness and negative effects on the environment. Therefore, in addition to the fossil fuel output, it is also necessary to pay attention to the measures on reduction of energy consumption and on the efficiency of its use.

The energy consumption in the Russian Federation is more than 3 times as large as that in the West European countries; the RF economy is one of the most energy-consuming economies in the world [1]. In this connection, the RF Government set in 2008 a rather ambitions goal: to reduce the energy consumption by not less than 40% (in comparison with 2007) by 2020. This goal has been formulated in the RF President Decree No. 889 dated 04.07.2008 "About some measures on improvement of energy efficiency and environmental safety of the RF economy"; the aforesaid decree has been put into the basis of the Federal Law No. 261 FZ (dated 23.11.2009, wording from 13.07.2015) "About the energy-saving and the improvement of energy efficiency as well as the introduction of changes into particular legislative documents in the RF".

In this document, the problem of reduction of energy consumption in construction industry and housing-and-communal services is considered to be one of the most important problems as the hot-water heating and hot-water supply require about 1/3 of the energy consumed, which may be reduced through the construction of energy-efficient buildings. The construction dynamics for residential buildings within the period from 1990 till 2014, presented in Figure 1, testifies to the fact that, since 2000, the residential floor areas became much larger, which also led to greater energy consumption.

[a] Corresponding author: Leonid.A.Shilov@gmail.com

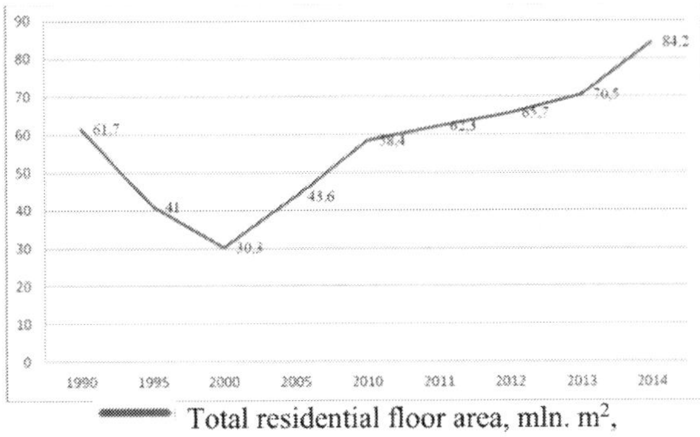

Fig. 1. The total residential floor area, mln. m^2, of buildings introduced into maintenance within the period from 1990 to 2014

The energy efficiency requirements for buildings and structures as well as the requirements to the determination of the energy efficiency categories are approved by the RF Government Decree No. 18 dated 25.01.2011.

The energy efficiency requirements for the buildings and structures are also included into the Federal Law No. 384-FZ dated 30.12.2009 (wording from 02.07.2013) "Engineering regulations concerning the safety of buildings and structures".

The question of energy-saving in construction was raised as early as in 1970s in connection with the concept of "sustainable development". The first energy-efficient buildings (the structures with low energy consumption or no heating facilities [2–4]) appeared in 1974, after the world energy crisis.

The first energy-efficient 7-storey office building with the total floor area of 16 350 m^2 has been designed by the architects N. Isaak and E. Isaak in the USA. The optimum planning and air flow distribution reduced the energy consumption for the ventilation system. The energy spent on the air cooling or air heating was also reduced due to the recuperating heat exchangers. The electric energy consumption was reduced due to the control over the illumination devices: the electric light intensity in the building depends on the natural light level [5].

In 1979, in Finland, the second energy-efficient building (ECONO-house) was put into service; the building was designed and erected under the supervision of the architect H. Kautonen. The structure provides for a rational use of the internal volume: the fencing structure surface is minimized and so the heat loss is reduced.

The trend used in the first energy-efficient building in the USA develops successfully even in the Northern countries, and the second trend, put into the basis of the Finnish energy-efficient house, started a new tendency in the construction industry called "Intellectual buildings".

In 1977, new energy-saving construction norms and standards were taken at the State level in Denmark (Danish BR77 Standard); and in 1980, those were taken in Sweden (SBN-80, Svensk Bygg Norm). By 1988, Sweden reduced the heat losses in residential buildings by almost 50%. Denmark consumed the heat energy in residential buildings in 1985 by 28% less in comparison with that in 1972.

Up to now, the Danish and Swedish energy consumption standards in construction are considered the most strict ones in the world.

In 1980s, the joint Swedish and German efforts created the concept of a passive building (Passivhaus), which has a continuous building shell with an improved heat insulation and the thermal conductivity of less than 0.15 $W/m^2 \cdot K$ [6].

The founder of the concept is doctor Feist, who calculated the energy balance of the buildings until he reached the parameters of the building which didn't require any special heating system.

In accordance with this concept, the first experimental Passivhaus-prototype was erected in Germany in 1991; a successful building service led to a series construction of such buildings: there were more than 7000 passive buildings in Germany by 2007.

The next stage of development of "passive buildings" is considered to be an "active building" which must provide the electric energy and hot water for itself. Nowadays typical equipment in the active building includes a solar collector for the production of hot water, a solar power station on the roof and a heat pump transforming the low potential heat of the earth or the domestic sewage into the hot water. Thus, a real "passive building" includes a power station as well.

The first active building has been erected in Denmark; first, it consumes not much energy (like a passive building); second, it produces the energy and can pass it to the central network, which is paid in several European countries.

The highest stage of the development of energy-efficient buildings is the "Sustainable building" construction. A sustainable building is a building with a comfortable microclimate in its rooms, where the energy of the nature is used to the greatest extent and the energy elements of the building as a whole are optimized.

Nowadays, the best energy efficient building is an office building in Norway which produces more energy (200 000 kW · h) than consumes. This unique "Kjørbo" building has the total floor area of 2600 m^2; it is situated at the Sandwik embankment (suburb of Oslo); after the global reconstruction it requires about 100 000 kW · h / year. The electric energy is generated through the solar panels. The tight joints of walls, floors, windows as well as a proper heat insulation provided for the minimum heat losses, and the façade facing considerably reduced the heating consumption of the building in summer [7].

In Russia we also have some examples of energy-efficient construction: two buildings are erected in Sevastopol for the purpose of the study of possible recommendations on the improvement of energy efficiency of buildings.

The first building is an administrative building, where some energy-saving measures are suggested: the walls are filled with aerated concrete blocks or with a local porous natural material (shell rock; its thermal conductivity is close to that of the aerated concrete). The glazing area of the Southern façade considerably exceeds that of the Northern facade; the glazing includes the energy-saving windows, etc.

The second building is a residential one; the same measures were taken there as well. In addition to that, the building contains a hanging facade with a ventilated gap; there are some mineral wool panels in the gap. Both the orientation of the building and the landscape were taken into consideration in the process of construction of the building.

A two-storey energy-efficient building is being under construction in Moscow Region. It is supposed that the building will consume not more than 45kW · h / (m^2 · year) due to the use of modern mineral wool insulation, the energy-saving multiple glass units, the air recuperation system and the heat pump.

In average, the heat energy consumption in low buildings equals 150 to 300 kW · h / (m^2 · year) (the normative value is 95 to 195 kW · h / (m^2 · year)).

In Tomsk, a block of flats has been included into the highest energy-efficiency category "A" due to a complex introduction of energy-saving technologies.

The structure of energy consumption for a block of flats before the reconstruction and after that is given in Table 1 [8].

Table 1. Main service parameters of a residential building: before the reconstruction and after that.

№	Energy resource	Basis month energy consumption	
		January, 2012	**January, 2014**
1.	Electric energy, kW · h	2509,00	2714,777
2.	Heat energy, Gcal	53,57	23,66
3.	Cold water, m^3	215,76	208,082
4.	Hot water, m^3	156,24	116,816
5.	Sewage, m^3	372	324,898

The plan of capital repair works may include energy-saving engineering solutions directed at a considerable improvement of the energy efficiency category of the building.

Since 2008, Moscow State University of Civil Engineering has an Energy Efficiency Situation Centre (EESC) created on the basis of the Research-and-Education Centre for Information Systems and Intellectual Automation in Construction.

Now the EESC carries out the energy-efficiency control over two university buildings with two total floor area of 18 000 m^2. The EESC performs an operative control over the engineering systems, the data analysis, the verification of engineering and technology solutions as well as of the software for the automated systems of control over buildings.

The EESC is an experimental site for a new approach in construction: the consideration of a construction project within the whole period of its life cycle: from the moment of erection to the time of its utilization. This approach suggests both the development of the concept and the design of construction projects with consideration of conditions of their further maintenance. Here it is possible to achieve a considerable saving in maintenance cost (cost of resources, services, possible repair works and works in hazardous situations) [9].

Therefore, the aforesaid examples are only single instances and do not contribute much to the development of energy-efficient construction.

The problem lies in the gap between the practice of the use of new construction materials and engineering equipment systems in buildings and the evolution of efficiency and even expediency of their use.

Here the most problematic are the facts that we do not have home-produced energy-efficient equipment and the energy-efficient engineering systems are rather expensive.

Conclusions

Today, city's sustainable development is an actual important task, which is to maintain the balance of city's development and the acceptable level of the environment.

Sustainable development of cities involves the creation of beautiful, healthy, energy-efficient city, ensuring full satisfaction of the population's needs.

Now the energy-efficient buildings require considerable additional investments in comparison with traditional buildings. The first stage of the process of improvement of energy efficiency are the heat-carrier regulation systems; it is necessary to install these systems in existing buildings, the pay-back period equals 5 to 6 years.

It is possible to use in Russia other measures and instruments which are successfully applied in energy-efficient buildings abroad.

It should be noted that in European countries the warmth-keeping layer may be made only 15 to 20 cm thick, while in Russia this layer should be made about 20–30 cm thick, the walls become thicker and the total floor area becomes less. So it is expedient to develop new construction materials, which are efficient, cheap and durable.

In addition to that, more attention should be paid to the problems of energy efficiency control over the construction projects during their life-cycle.

References

1. L.A. Shilov, A.O. Adamtsevich, L.A. Shilova, IRR, **5** (**147**), 27–31 (2015)
2. I.V. Kryaklina, E.V. Sheshunova, I.L. Grek, E, **1** (2014)
3. N.V. Grinkrug, S.A. Kostrikov, IJABR, **5-3**, 389–394 (2015)
4. I.M. Mukhutdinov, A.G. Orlov, V.V. Kondrashin, MNIZH, **5-1** (**12**), 84 (2013)
5. Information source: http://www.c-o-k.ru/review/samye-izvestnye-v-mire-energoeffektivnye-zdaniya
6. M. Endhardt, E, **5**, 20–24 (2010)
7. Information source: http://greenevolution.ru/2014/05/15/samoe-energoeffektivnoe-zdanie-v-mire
8. A.A. Nechepurenko, E, **3** (**34**), 48–50 (2014)
9. A.A. Volkov, P.D. Chelyshkov, A.V. Sedov, VE, **5** (2015)

E3S Web of Conferences **6**, 03005 (2016) DOI: 10.1051/e3sconf/20160603005
ICSC 2016

Model of Eco-Industrial Park Development as a Tool for Fostering Energy Efficient Economy

Evgeniy Tikhanov[1a], *Vadim* Krivorotov[1], *Aleksey* Kalina[1], and *Sergei* Erypalov[1]

[1] Ural Federal University named after the first President of Russia B.N. Yeltsin, 620002, Yekaterinburg, Russia

Abstract. Energy efficiency plays a key role in the development of modern socio-economic systems, such as cities. Energy efficiency can be achieved through creation and development of eco-industrial parks. This article describes the concept of such parks and their peculiar characteristics and provides an overview of research literature on their development, energy efficiency and competitiveness of their resident enterprises. Performance of eco-industrial parks can be evaluated by applying a set of indicators, which demonstrate the competitiveness of park residents. This system of indicators can be used to model the management of eco-industrial park development. This article also describes a methodological approach to evaluate competitiveness of resident enterprises: the integral index of competitiveness, which synthesizes their current and prospective competitiveness. Comparative analysis (conducted by applying 'PATTERN' method, Planning Assistance Through Technical Evaluation of Relevance Numbers) also takes into account specific indicators of residents' competitiveness. This research uses the methods of static modelling to create an algorithm for building a model which establishes the correlation between the indicators of the park's performance and the level of its residents' competitiveness. The proposed model was applied to develop a program of management for eco-industrial park 'Tagil'.

1 Introduction

Nowadays developed countries are going through a new wave of industrialization, which is driven by knowledge-intensive innovations and advanced technologies in the humanities and social studies. Such industrialization significantly raises the energy efficiency standards and brings urban space to a totally new level. Thus, energy efficiency encourages qualitative economic growth, gives the competitive edge to the existing and newly established manufacturing enterprises and provides long-term sustainable development of urban areas. Cities and their infrastructure are becoming more efficient, safe, and comfortable.

The key areas of enhancing economic energy efficiency include stimulating the growth of enterprises with low energy intensity; improvement of the infrastructure; cost saving at the stages of production, transportation and consumption of energy. Industrial parks productively combine the above-mentioned measures and are considered as one of the most

[a] Corresponding author: tjohn90@mail.ru

efficient forms of production organization in the world. Industrial parks are generally understood as centrally managed areas with well-organized infrastructure, which are intended for establishing new production enterprises. At present there are over twenty thousand industrial parks in over ninety countries of the world.

In many countries of Europe, North America and Asia, industrial parks are seen not only as a tool for attracting private investment, creating new workplaces, increasing tax revenues and diversification of economy but also as means of developing energy efficient economy, implementation of innovations and advanced processing of raw materials.

One of the most promising concepts developed in the recent decades is that of eco-industrial parks (EIPs). The main difference between EIPs and their industrial counterparts is that management companies follow comprehensive strategies of development, which, apart from improving the performance of resident enterprises, seek to reduce their negative environmental impact. EIPs are expected to meet the following goals: to form a balanced infrastructure; to increase the efficiency of energy and other resources; to create resource-saving and low waste technologies; to establish partnership relations between enterprises; to curb the environmental pollution; to organize advanced processing of waste and encourage resident companies to reuse waste.

2 Modern research on eco-industrial parks, energy efficiency and competitiveness of enterprises

The concept of EIPs emerged in the USA in the early 1990s. It was intended to bring together producers of commodities and services in order to improve their economic performance and at the same time to protect the environment by managing natural resources collectively. Residents of EIPs uphold high standards of energy efficiency and engineering solutions and work together to ensure that the waste of one company provides materials for another while processed materials are used to manufacture the end product. Nowadays such EIPs are successfully developing in such countries as Canada (Hinton); the USA (Reventure Park); Denmark (Kalundborg); Finland (Rantasalmi); China (Dailan, Suzhou, Nanhai); Indonesia (Semerang, Industri Sona Maris); Japan; India; Vietnam; Thailand; and Mexico.

Most specialists consider EIPs as a tool to develop energy efficient economy and as an effective mechanism of providing resident enterprises with a competitive advantage [1–5]. Therefore, it is important to develop a universal model for management of EIPs' development. The key goals in this process should be to make EIPs more energy efficient and to improve their residents' competitiveness.

Competition is the main driving force of development in market business models and the relevant research literature provides a wide range of approaches to competitiveness evaluation. The existing approaches can be divided into the following groups:

1. Competitive strategy methods: for example, the matrix of the Boston Consulting Group 'Market Share'/'Market Growth' [6]; McKinsey's matrix 'Competitive Strength of a Business Unit/Industry Attractiveness' [7]; M.E. Porter's matrix 'Strategic Objective/Strategic Advantage' [8];
2. Methods based on the value chain concept. This approach was first introduced by M. Porter [9] and then it was developed by a number of other famous researchers such as G. Gereffi [10] and J. Humphrey [11];
3. Approaches and methods which take into account competitiveness of production [12, 13 and others];
4. Methods based on estimation of the business value [14, 15 and others];
5. Methods which are based on the game theory and focus on the strategies of choosing optimal behaviour of business entities [16–19 and others];

6. Integrated methods.

These days, however, there is no single approach to evaluation of competitiveness or energy efficiency. As a rule, the latter is evaluated according to a set of indicators, which are determined individually for each specific case. In general, we can identify two main approaches to selecting such systems of indicators.

The first approach divides all energy efficiency indicators into economic (value) indicators; technical and economic (physical) indicators; and the indicators related to how intensively companies use energy efficient technologies [20, 21 and others].

The second approach classifies energy efficiency indicators according to types of business activity: it deals with specific industries and with the combinations of different types of activities inside these industries as well as their energy intensity. Such approach is applied, for instance, in the methodology of Asia Pacific Research Center [22]; in the projects of the International Energy Agency [23, 24]; the French Environment and Energy Management Agency (ADEME) [25] and many others.

Another method which is widely applied for evaluating energy efficiency is data envelopment analysis. It is based on comparative analysis of complex technical, economic and social systems. It was introduced by M.J. Farrell [26] and developed by A. Charnes, W.W. Cooper, R.D. Banker [27, 28], and other specialists. It is also applied by Yi-Ming Weia [29], K. Mukherjee [30], and B. Ang [31].

3 System of indicators characterizing competitive growth of resident enterprises

To systematize the advantages gained by residents of EIPs the factors of their success can be divided into the three major groups:

1. 'Internal' factors, that is, conditions enhancing the creation and further development of energy efficient enterprises. This group of factors include technology intensive infrastructure; organization of material and energy flows between the residents; open access to information about the waste production, the needs of the EIP's residents in raw materials and the energy they use;
2. Local factors, that is, geographical and logistical advantages of the host area, available human, material, and energy resources;
3. Regional factors, that is, the advantages of locating resident enterprises on the territory of a specific country or region. These include such factors as state support and a favourable investment climate.

We can use a system of quantitative indicators to evaluate and analyze the impact of the above-described factors on competitiveness of enterprises. The analysis of the indicators characterizing the performance of an EIP allows us to develop a complex of management measures to enhance its growth and to create a program of its development by taking into account the real needs of the existing residents.

These indicators can be divided into the two major groups:

1. The indicators which correspond to the 'internal' factors of an EIP's energy efficiency;
2. The indicators which correspond to the external factors (local and regional).

The system of indicators proposed in this paper is shown in Figure 1 and includes nine groups of indicators:

1. Eco- and energy efficiency. Resident enterprises should be using cutting-edge energy saving and environmentally friendly technologies and form closed non-waste production cycles within the framework of a single manufacturing chain or similar kinds of activity. All this would enable them to optimize their industrial processes, enhance their resource efficiency, save the waste disposal costs and the costs of procuring the necessary materials.

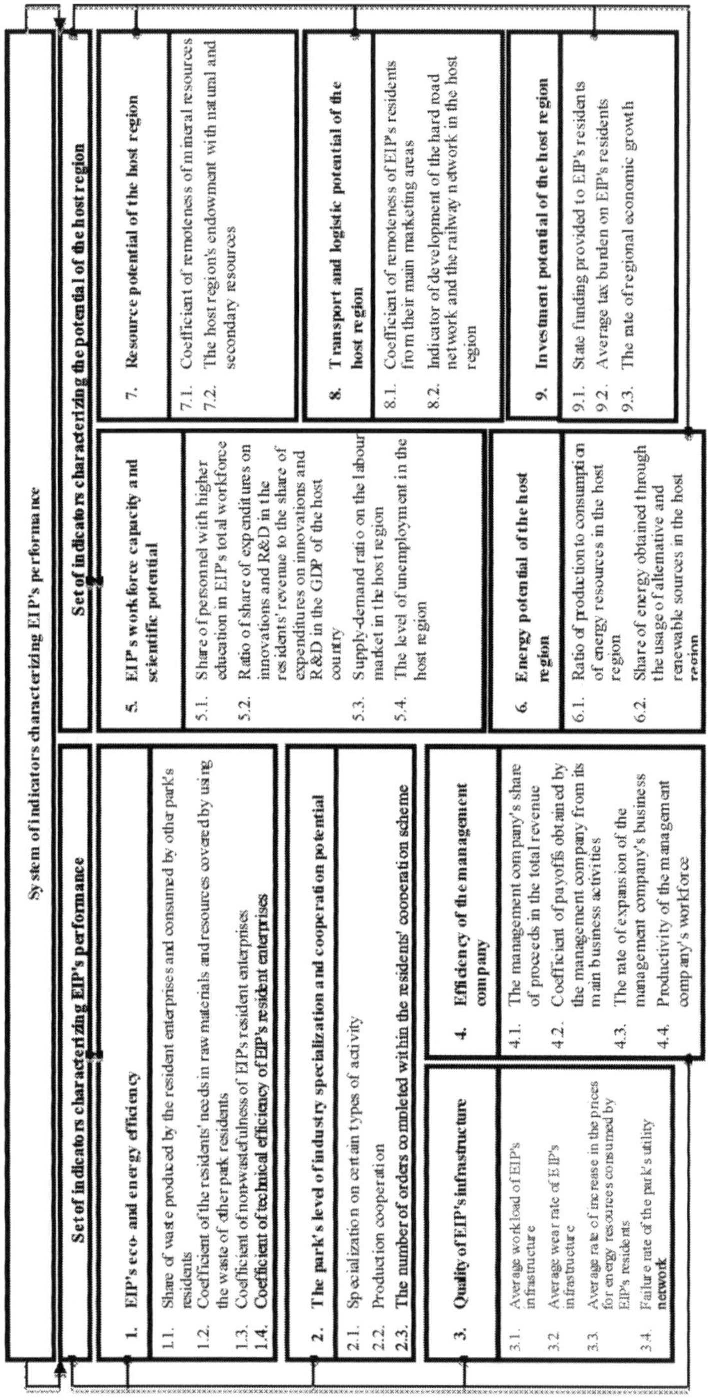

Figure 1. System of indicators characterizing EIP's performance

2. Industry specialization and cooperation potential. These are indicators which reflect the potential of the park and its residents regarding their ability to establish long-term economic relations and conduct joint or technologically related production.

3. The quality of the park's infrastructure. These indicators characterize the park's key asset: its engineering, transport and manufacturing infrastructure. They determine time and

financial expenditures of the residents at the initial stage of the park's development and significantly affect their level of operational expenditures.

4. Efficiency of the park's management company. This group of indicators characterizes the quality of the management company's services provided according to the 'single window' principle, which is of paramount importance for ensuring competitiveness of resident enterprises.

5. Workforce capacity and scientific potential of the host region. These indicators characterize the available workforce, including highly qualified personnel; the quality of the regional educational system; innovative and intellectual potential of the region.

6. Energy potential of the host region. This group of indicators reflects the amount of available energy resources; the availability of alternative and renewable energy sources (for example, wind power generation, hydropower industry, and solar cell panels).

7. Resource potential of the host region. These indicators characterize the availability of natural and secondary mineral resources in the host region, which can be used by resident companies and determine the industrial specialization of this region.

8. Logistical potential of the host region. These indicators characterize the location of the park in relation to the available transport network. The quality of the transport network has a significant impact on the performance of resident companies.

9. Investment potential of the host region. These indicators correspond to the investment climate of the host region, that is, the environment for business and the tools for supporting investment activities.

The above-described system of indicators allows management companies to conduct a comprehensive analysis of their investment sites, evaluating their key advantages and drawbacks. Management companies can also compare specific indicators of the park in question with those of its competitors and thus identify the areas for further improvement. Moreover, this system can be used to elaborate an EIP development program aimed at enhancing the residents' competitiveness.

4 Methodological approach to evaluation of the resident enterprises' competitiveness

Another element of the proposed model is the methodological approach to evaluation of the resident enterprises' competitiveness. These authors suggest considering competitiveness as a synthesis of current and prospective competitiveness.

Current competitiveness comprises the energy efficiency the enterprise has already achieved and its financial stability.

Prospective competitiveness (competitive potential) corresponds to the enterprise's capability of maintaining and increasing its competitiveness in the future taking into account its actual dynamics and the degree of risk attached to this enterprise's activity.

In order to link these two values together in one indicator, we need to apply the geometric mean formula (Formula 1) (this method has already been applied by these authors to solve similar problems [32, 33]):

$$K = \sqrt{K_C \times K_P}\ , \qquad (1)$$

where K is the integrated index of competitiveness;
K_C is the indicator of the enterprise's current competitiveness;
K_P is the indicator of its prospective competitiveness.

To evaluate competitiveness of a business unit we need to compare the relevant indicators of its activity with those of its competitors, which are assumed as the basic or reference model (Formula 2):

$$K_i = \frac{C_i}{C_{basic,i}}, \quad (2)$$

where K_i is the value of the i indicator of the company's competitiveness;
C_i is the value of the i coefficient of the company's competitiveness;
$C_{basic,i}$ is the basic (reference) value of the i coefficient of the company's competitiveness.

Depending on the research objectives, basic indicators can be the indicators characterizing the activity of the main competitor, a group of competing companies, industry-average indicators in this region, country, or a group of countries. When selecting competing companies for the analysis, it is crucial to take into account the types of their business activity and their product range; the territory's specific features; and the range of the companies' activity.

Formula 2 can be interpreted in the following way: the higher is the value of the coefficient, the more competitive is this enterprise in this indicator as compared with its competitors.

Current competitiveness is determined by the indicators of efficiency and financial sustainability (Formula 3) while prospective competitiveness, by the indicators of the enterprise's potential and risk-taking (Formula 4):

$$K_C = \sqrt{K_E \times K_{FS}}, \quad (3)$$

where K_T is the current competitiveness of the enterprise;
K_E is the indicator of its efficiency;
K_{FS} is the indicator of its financial sustainability.

$$K_P = \sqrt{K_{EP} \times K_R}, \quad (4)$$

K_P is the prospective competitiveness of the enterprise;
K_{EP} is the indicator of the enterprise's potential;
K_R is the indicator of the enterprise's level of risk-taking.

Each of the above-described indicators, in its turn, includes three initial estimate indicators. Therefore, the system of indicators characterizing efficiency of enterprises comprises twelve initial indicators, which are then divided into four groups corresponding to the indicators of the companies' current and prospective competitiveness.

5 Algorithm of building a model of correlation between the indicators of the park's efficiency and the level of competitiveness of its residents

The acquired quantitative evaluation of the key indicators of the park's efficiency and the indicators of its residents' competitiveness in several retrospective periods enables us to establish the mathematical correlation between the specified values. This will become the basis for further modelling. It is important to mention that to make this model suitable for practical application the following requirements must be met:

1. Functional relationship between competitiveness of the EIP's residents and the indicators of the site's performance can be determined by using widely applied mathematical tools.
2. The starting data must be representative, that is, it should cover a significant period of time.

3. Final evaluation should include not only the evaluation of specific residents but of the whole group of core residents. This will provide a more objective picture for further development of measures enhancing competitiveness of the largest enterprises.

The model is built according to the following algorithm:

1. Collection of data on a number of consecutive periods. To make the data more representative the calculations of the relevant indicators should cover not less than five equal time periods.

2. Establishing the correlation between the indicator of competitiveness of the core group with each of the indicators of the EIP's performance. To model this relationship we applied the method of correlation and regression analysis. The initial system of models characterizes the correlation between competitiveness of the core group of residents and each of the indicators of the park's efficiency. This system looks the following way:

$$R_i = f(X_{i1}, X_{i2}, \dots, X_{in}), \tag{5}$$

where R_i is the indicator of the core residents' competitiveness, calculated in accordance with the values of the indicators of the i set of indicators;
$X_{i1}, X_{i2}, \dots, X_{in}$ are the indicators of the i set of indicators.

At this stage we need to choose the mathematical function which would determine the cause-and-effect relationship between the indicators in question and the parameters of this function related to the intensity of this relationship and its direction.

3. To estimate the accuracy of this evaluation we need to analyze the coefficient of regression, the intensity of the relationship, approximation errors and other parameters and to determine the most significant indicators of the park's performance.

4. Exclusion of unreliable models and models contradicting the logic of the process in question.

5. Building the final model, which illustrates the correlation between the core residents' competitiveness and the indicators of the park's efficiency and which excludes the effect of multicollinearity of the explanatory factors. The acquired model can serve as a basic tool to develop a complex of management measures to improve the EIP's performance.

6 Practical testing of the model for the development of an eco-industrial park

The proposed model was used to solve a practical task of developing a program for the development of EIP 'Tagil', which is located in Sverdlovsk region (Russia). The project includes horizontally and vertically integrated chemical production companies, which seek to meet high standards of environmental safety and are located in close proximity to raw materials, oil and gas, energy sources, human resources, and diverse market outlets. The correlation between the competitiveness level of the core residents of 'Tagil' and the indicators of this park's performance looks the following way:

$$R = -0{,}58 + 0{,}14 \times X_1 + 0{,}31 \times X_2 + (-0{,}08 \times X_3) + 0{,}22 \times X_4 + 0{,}03 \times X_5, \tag{6}$$

where X_1 is the non-wastefulness coefficient of the park's residents;
X_2 is the level of cooperation achieved on the territory of this park;
X_3 is the coefficient of remoteness of the park's residents from mineral resources;
X_4 is the state support provided to the park's residents;
X_5 is the share of energy obtained through the use of alternative and renewable sources on the park's territory.

Analysis of this correlation shows that the key factors of residents' competitiveness are the following: environmental safety and non-wastefulness of production; close cooperation between the enterprises operating within the framework of a single technological process; and the state funding received by the residents. Therefore, the development of EIPs should prioritize the following strategies:

1. Raising the standards of ecological and energy efficiency of the park's resident enterprises;
2. Enhancing and expanding interfirm cooperation in production, research and marketing;
3. Assistance to residents provided by the management company in order to attract financial and non-financial state support for the development of production on the territory of park 'Tagil'.

7 Conclusion

The proposed system of management of the EIP development is targeted at enhancing competitiveness of its residents. It is a practical tool, which can be used by management companies to elaborate a program of the park's development and create favourable conditions for the growth of resident enterprises. Sustainable development of EIPs will contribute to boosting the local and regional energy efficient economy.

Acknowledgements

This research was supported by the Government of the Russian Federation (Act 211, agreement № 02.A03.21.0006).

References

1. A.J. Lambert, F.A. Boons, Technovation, **8**, 471–480 (2002)
2. M. Dwortzan. The Greening of Industrial Parks. Technology review, **1**, 18–19 (1998)
3. Z. Wang, D. Sun, L. Shi, D. Hu, Y. Xu, Annals of the New York Academy of Sciences, **1**, 145–153 (2010)
4. D. Emerson, In Business, **3**, 14–20 (2003)
5. L. Metzler, M. Lechner, T. Hayes, American City & County, **1**, 14–19 (2002)
6. B. Henderson, The Product Portfolio, https://www.bcgperspectives.com/content/Classics/strategy_the_product_portfolio
7. McKinsey & Company (2008), http://www.mckinsey.com/insights/strategy/enduring_ideas_the_ge_and_mckinsey_nine-box_matrix
8. M.E. Porter, *Competitive Strategy: Techniques for Analyzing Industries and Competitors* (Free Press, 1998)
9. M.E. Porter, *On Competition* (Boston, 1998)
10. G. Gereffi, *A Commodity Chains Framework in Analyzing Global Industries* (Durham, 1999)
11. J. Humphrey, *Upgrading in Global Value Chains*: ILO Working Paper No. 28 (2004)
12. L.N. Chaynikova, V.N. Chaynikov, *Competitiveness of the Product* (Tambov, 2007)
13. R.A. Fatkhutdinov, *Strategic Marketing* (Moscow, 2000)
14. T. Copeland, T. Koller, J. Murrin, *Valuation: Measuring and Managing the Value of Companies* (New-York:, 1995)
15. M.C. Scott, V*alue Drivers. The Manager`s Framework for Identifying the Drivers of Corporate Value Creation* (1998)

16. V. Cerasi, B. Chizzolini, M. Ivaldi, Applied Economics, **34**, 2213-2225 (2002)
17. B. Doberman, H. Gatingnon, G. Sargsyan, *Using Attraction Models for Competitive Optimization: Pitfalls to Avoid and Conditions to Check*: INSEAD Working Papers No. 27 (2006)
18. V. Kadiyali, K. Sudhir, R. Vithala, International Journal of Research in Marketing, **18**, 161–186 (2001)
19. S. Sriram, V. Kadiyali. *Channel Responses to Brand Introductions: An Empirical Investigation* (2007)
20. B.W. Ang, S.Y. Lee, Energy Economics, **16**, 83–92 (1994)
21. B.W. Ang, K.H. Choi, The Energy Journal, **18** (3), 59–73 (1997)
22. Energy Efficiency Indicators: A study of Energy Efficiency Indicators for Industry in APEC Economies, http://aperc.ieej.or.jp/file/2010/9/26/Energy_Efficiency_Indicators_for_Industry_2000.pdf
23. Energy Indicators System: Index Construction Methodology, http://www1.eere.energy.gov/ba/pba/intensityindicators
24. Energy Policies of IEA Countries, http://www.iea.org/publications/freepublications/publication/canada2009.pdf.
25. Indicators to Measure the Contribution of Energy Efficiency and Renewables to the Lisbon Targets, http://www.isi.fraunhofer.de/isi-en/profile/publikationen.php
26. M.J. Farrell, Journal of the Royal Statistical Society, **120**, 253–281 (1957)
27. A. Charnes, W.W. Cooper, E. Rhodes, European Journal of Operation Research, **2** (6), 429–444 (1978)
28. R.D. Banker, A. Charnes, W.W. Cooper, Management Science, **30** (9), 1078–1092 (1984)
29. Y-M.Weia, H. Liao, Y. Fan, Energy, **32** (12), 2262–2270 (2007)
30. K. Mukherjee, Energy Economics, **30**, 76–96 (2008)
31. P. Zhou, B.W. Ang, Energy Policy, **36**, 2911–2916 (2008)
32. V. Krivorotov, A. Kalina, Zh. Belyaeva, S. Erypalov, Proceedings 7th EuroMed Conference of the EuroMed Academy of Business, 827–835 (2014)
33. V. Krivorotov, A. Kalina, S. Erypalov, Zh. Belyaeva, Competitiveness Review, **26** (2) (2016)

Construction and optimization of a power complex with a distributed generation on the basis of renewables and methods of artificial intelligence (on the example of the Republic of Tajikistan)

Vadim Manusov[1], *Alifbek* Kirgizov[2], and *Javod* Ahyoev[1a]

[1]Novosibirsk state technical university, 630073, Novosibirsk, Russia
[2]Tajik technical university of a name of the academician M.S. Osimi, 734042, Dushanbe, Republic of Tajikistan

Abstract. In this article the basic principles and classifications of small hydroelectric power stations depending on the power of system in which they work are provided. Especially, in the conditions of Tajikistan when all small hydroelectric power stations in some regions carry out the functions of big hydroelectric power stations. We suggest to use new classification by using function of fuzzy logic. The new concept of a power complex (HUB) on the basis of the renewed energy source (RES) is also offered.

1 Introduction

In recent years the agricultural enterprises and rural population of the Republic of Tajikistan has been facing constant deficiency of electric energy which is the main source for heating in households. Today this deficiency especially became aggravated and the population in the majority of rural areas receive the electric power during the winter time only few hours per day. It leads to unpleasant consequences in all spheres of life and agricultural activity of areas, namely: in economy, education, health care, the social sphere.

2 Relevance

Rural, especially remote and hard to access regions of the Republic of Tajikistan can be reliably provided with the electric power only due to construction of small hydroelectric power stations (MGES) or other renewables, such as wind turbines and solar batteries. Potential resources of a small-scale power generation of the Republic of Tajikistan in general are very great and significantly over exceeds the real demand, even with considering the perspective development of power engineering sphere in the long term future.

[a] Corresponding author: Javod_66@mail.ru

The availability of water flow, temperature, and wind speed is required for the production of electricity, that is for determining the optimal installed capacity and an average annual energy production by small hydro power plants and wind power plants.

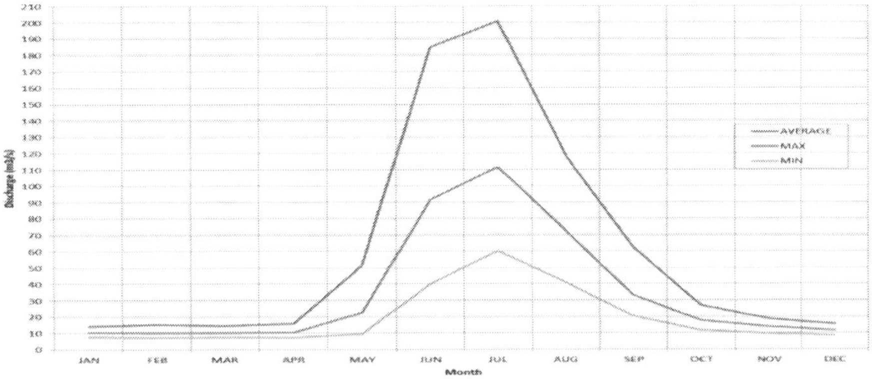

Fig. 1. A typical graph of the water regime changes in small water inflows of the Eastern Pamirs.

Average discharge highly varies from year to year, as shown in Figure 1, from 24 m^3/s to 48 m^3/s from average annual discharge of water.

Despite all this, progress in implementation of the program of development of the small-scale power distributed generation and construction of Small Hydro Power Stations (SHPS) in the Republic of Tajikistan is very modest. One of the main reasons for such slow development is lack of accurately developed strategy for development of the small distributed hydropower of the Republic and the main concept of development of a power complex which in literature is characterized now by the term HUB. HUB is a power knot (complex) which includes different types of power sources such as mineral raw materials, for example, coal, oil products, etc. HUB offered in this work represents composition of the distributed renewable sources of a net energy of different types, such as small hydroelectric power stations, low-power wind turbines, solar batteries and static stores of energy (BESS – Battery Energy Storage System). This is the fundamental difference of the offered HUB in the Republic of Tajikistan.

Problem definition

Currently, there is confusion in the concepts of small-scale power generation. Along with the term "small hydropower" also "nonconventional", "alternative" terms are used that creates confusion and difficulties in the use of world experience of the developed countries which have different condition in comparison to the Republic of Tajikistan.

First of all, it is required to concretize concept of small hydroelectric power station. Specification of classifications of Small Hydro Power Stations (SHPS) by power and local electric networks which will participate in work of HUB. Especially it concerns the region of Gorno Badakhshan Autonomous Oblast (GBAO) where all kind of basic hydroelectric power stations (both existing and planned), formally classified as "small", however carrying out actually the functions of "big" power station. It can create uncertainty in their status, ownership and the role of the state in their management.

Nowadays, not explicitly, but there is an understanding that small hydroelectric power stations, are those large hydroelectric power stations, but in a small size. Apparently, the division is necessary not only according to capacity, but also according to structure of consumers, operating mode, meeting the production schedule, construction and connecting consumers is necessary. It would be correct in order to define the limit of the SHPS capacities to link relative values of power to fixed capacity of the energy system to the extent of the development of the country and other factors. It would be necessary to connect limits of capacities of SHPS with development rate of the country, one of the indicators is the total capacity of a power supply system. The accepted classification parameters for big hydroelectric power stations and SHPS are almost identical, except the capacity limit which in the different countries is applied differently. It creates additional difficulty for definition of exact classification of SHPS especially in low-capacity systems. More exact and similar classification can be determined by function of belonging of station to small, as SHPS capacity is not fixed value, but intersected classes.

The histogram of distribution of the existing hydroelectric power stations by capacity is presented in figure 1, from total number of 325 hydroelectric power stations of the Republic of Tajikistan.

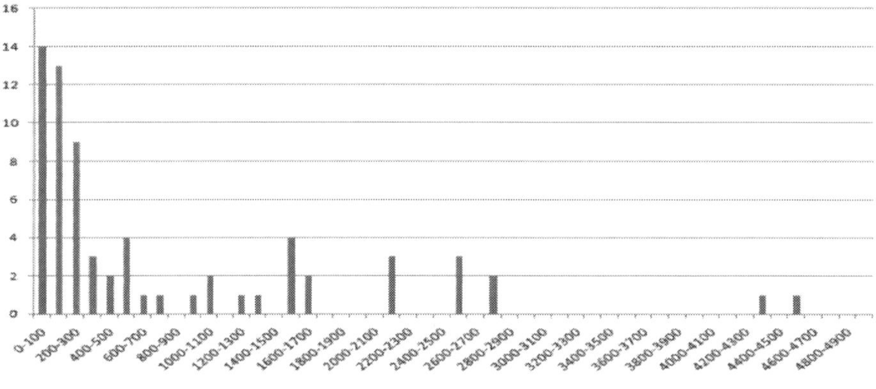

Fig. 2. Histogram of power -stations of Tajikistan by capacity.

Based on their functions of participation in the wholesale market of the electric power, hardware and types of management it is expedient to highlight the following groups of distributed generation of hydroelectric power station: micro hydroelectric power station, mini hydroelectric power station and small hydroelectric power stations. Since clear breakdown of hydroelectric power station by capacity depending on their quantity isn't possible, indistinct classification on the basis of functions of belonging of intersected classes is accepted which is presented in figure 2.

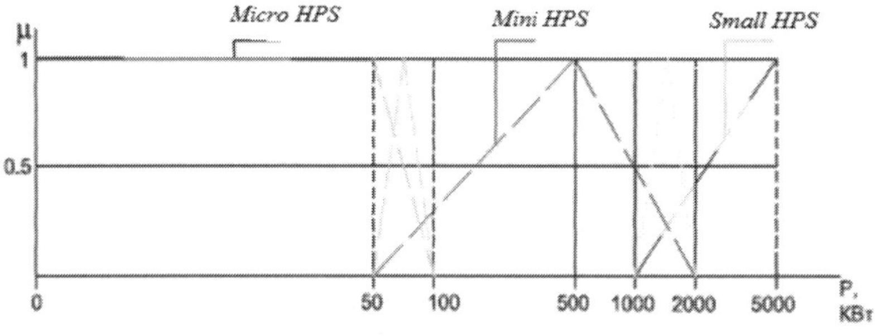

Fig. 3. Fuzzy classification of SHPS adapted to conditions in the Republic of Tajikistan.

In Figure 2 it is visible that micro hydroelectric power station includes stations from 0 to 50 kW, the mini hydroelectric group includes stations with the capacity from 50 to 2000 kW, and small hydroelectric power station includes the stations with the capacity from 1000 to 5000 KW.

This classification ascribes some degree of overlap between micro and mini HPS in the range of 50 to 100 kW and a fuzzy zone between mini and small HPS in range of 1000 to 2000 kW. Such a fuzzy classification is proposed for the first time and, evidently, it reflects the situation on the ground more realistically.

The mathematical expression for the functions of belonging in the general form of triangular L-R function is represented by the formula (1). For the classification of micro Hydro Power Plants (HPPs) the function of belonging is defined by the formula (2). For the classification of mini HPPs the function of belonging is defined by the formula (3). For the classification of small HPPs the function of belonging is defined by the formula (4)

$$\mu(x_i), \begin{cases} 0., c_i \leq x \leq a_i \\ \frac{x-a_i}{b_i-a_i}, a_i \leq x \leq b_i \\ \frac{c_i-x}{c_i-b_i}, b_i \leq x \leq c_i, \end{cases} \tag{1}$$

$$\mu(x_i), \begin{cases} 0., 0 \leq x \leq 50 \\ \frac{100-x}{c_i-50}, 50 \leq x \leq 100 \\ 0,, 100 \leq x, \end{cases} \tag{2}$$

$$\mu(x_i), \begin{cases} 0., 50 \leq x \leq 2000 \\ \frac{x-500}{500-50}, 50 \leq x \leq 500 \\ \frac{2000-x}{2000-500},, 500 \leq x \leq 2000, \end{cases} \tag{3}$$

$$\mu(x_i), \begin{cases} 0., 1000 \leq x \\ \frac{x-1000}{5000-1000}, 1000 \leq x \leq 5000 \\ 1, 500 \leq x \leq 2000, \end{cases} \tag{4}$$

Similar classification for the distributed generation can be executed for small mini and micro wind-power turbine in a class from 50 to 500 kW.

As the main development energy in the Republic is the share of RES and depending on their unpredictability and unevenness of transition into daily mode, their connection to HUB is the most rational way of increasing their reliable functioning.

One way to use the existing infrastructures more effectively they need to be considered as one integrated system in planning and operation of energy. For this purpose, it is necessary to connect all sources to one center which, allows optimization of their work depending on requirements and availability of energy.

If infrastructures are integrated properly, energy can be exchanged among them, integrated, in particular during congestion it is possible to exercise their control. For example, in certain times the transmission lines of energy can be overloaded and transfer can be reduced, having moved part of power energy to another a network and moving it to less critical points in system. But it would demand the integration of appropriate tools of all energy carriers and their collaboration.

Consideration of multiple inputs, the power center (HUB) can be used to meet the demand for consumption of production and clearly this center usually increases availability of energy to consumers because it doesn't depend on only infrastructure any more. The outlined concept is presented in Figure 4.

 DOI: 10.1051/e3sconf/20160603006

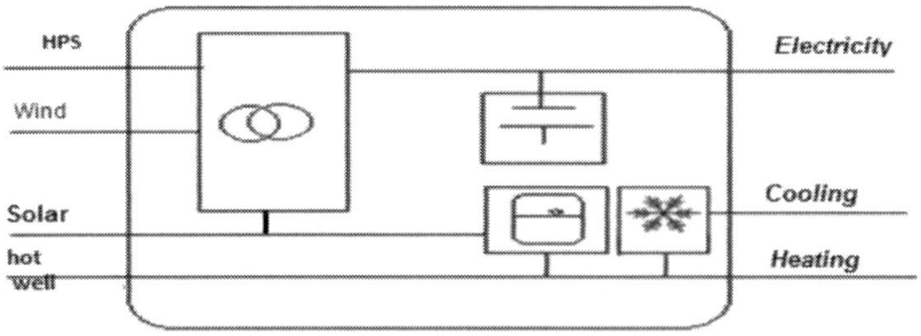

Fig. 4. Association of all sources in the uniform power center.

Along with hydropower resources in this work a study and evaluation of the energy potential of wind energy was conducted for different regions of the Republic of Tajikistan which is presented in Figure 5 with a buffer accumulation of energy. While assessing wind energy capacity in the cities of the Republic of Tajikistan, it was found out that the most common wind power units (WPU) are from 10 to 30 kW, although in some cases up to 120 kW. In accordance with the previously mentioned classification of the function of belonging the wind power units should be classified as micro stations (as defined by the formula (2). Therefore, the potential of wind energy is not very large, but it is useful for rural areas, as it is distributed generation.

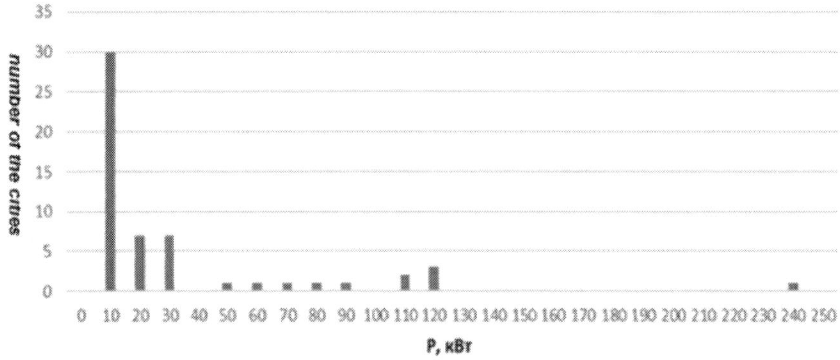

Fig. 5. The histogram for distribution of power wind turbines by cities of Republic of Tajikistan.

In addition to this, an assessment of solar energy potential was conducted, namely the clarity of days in the year. The daily solar radiation in the form of radiation kWh/m^2 per day. The average capacity by cities is from 1300 to 1600 kW per solar plant area of 1000 m^2. The mathematical expectation is 1400 kW. Figure 6 shows the distribution of solar power sources.

Installation of helio megawatt-class solar plants is considered. Per their capacity and classification of renewable energy plants, they belong to the class of mini power plants defined by the function of belonging according to the formula (3). Thus, the solar potential of the Republic of Tajikistan is considerably higher than wind power.

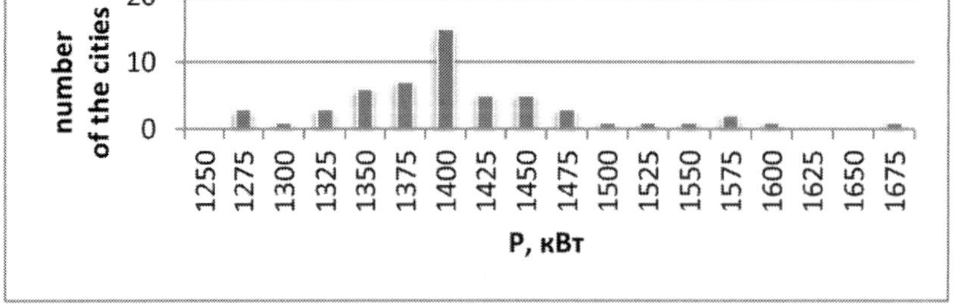

Fig. 6. Histogram of distribution of power of solar energy in the Republic of Tajikistan.

In the context of the power centers, the whole infrastructure of power supply it can be considered as system of the related power centers. In Figure 3 communication of three power centers is presented which includes integration of electric and thermal networks. It is an example for power supply of settlements and cities, which approximately divided into three areas: industrial enterprises, commercial spheres, and private houses.

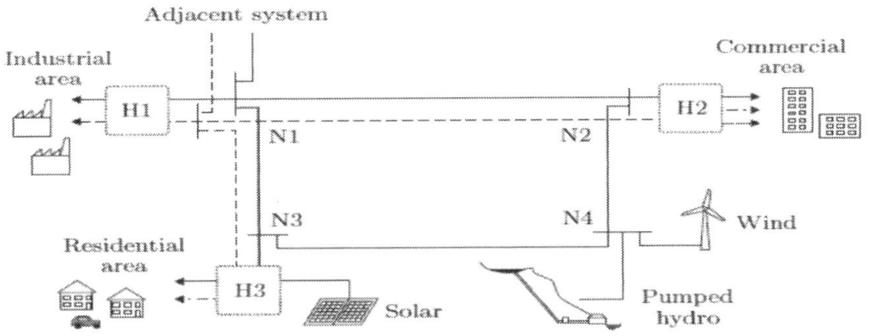

Fig. 7. Systems of related power centers.

Each area connects to distribution electric networks and thermal sources through the power centers. The internal arrangement of the centers is adapted for certain requirements of electric and thermal consumers. Power is transmitted through adjacent networks, photovoltaic sources, connected to H3 center, and wind-powered turbines and hydroelectric power stations are connected to an electric network through the N4 hub [1, 2]. This network can represent more remote hydroelectric power station near settlements and the cities where there are reservoirs for accumulation of energy of water or accumulators of energy of other type of BESS.

Transmission of energy between centers in the combined system for different types of energy carriers in the distributed system has to possess static stability that is subject to additional research [3]. Autonomous management of renewable energy sources, especially wind turbines, considered in [4, 5] are subject to further discussion.

Conclusion

The principles of creation of HUB offered in this work represents composition of distributed renewable sources of "net energy" of different types, such as small hydroelectric power stations, winder-powered turbines and low-powered wind stations, solar batteries and static stores of energy (BESS – Battery Energy Storage System). This is what is the fundamental difference of the offered HUB in the Republic of Tajikistan. The uniqueness of this HUB is in combined modes of operation of hydro, wind and solar resources in the simultaneous mode of competition with eacho the rand also complement each other Uniqueness of this HUB is reflected in a combined operating modes of water resources, wind and solar resources in the modes of the simultaneous competition among themselves and complementarity to each other. This way, electric HUB of the Republic of Tajikistan will work with the 100 percentage generation of a net energy of all types of renewable sources. Perhaps, it will be unique HUB in world power.

References

1. Z. Yuldashev, R. Yuldashev et al., Bulletin of the Tajik technical university, **2** (**26**). 59–61 (Dushanbe, 2014).
2. S. Rasulov, A. Kirgizov, and U. Kasymov, *Power: Ecology, Reliability, Safety: materials of the eighteenth All-Russian scientific and technical conference*. 99–101 (Tomsk, 2012).
3. M. Geidl, and G. Andersson, IEEE Transactions on Power Systems, Vol. 22, **1**, 145–155 (February 2007).
4. V.Z. Manusov, S.N. Udalov, *Modeling of wind power installations and management of them on the basis of fuzzy logic*: monograph (Novosibirsk, 2013).
5. V.Z. Manusov, S.K. Khaldarov, Applied Mechanics and Materials. Vol. 698: Electrical Engineering, Energy, Mechanical Engineering, EEM 2014, www.scientific.net/AMM.698.168

The model of creation of energy-saving competence of students of high school

Evald Zeer[1], *Elvira* Symaniuk[2], *Anna* Pecherkina[2], *Anastasia* Stepanova[2a], and *Nina* Symaniuk[3]

[1]Russian State Professional-Pedagogical University, 620012, Yekaterinburg, Russia
[2]Ural Federal University named after the first President of Russia B.N. Yeltsin, 620002, Yekaterinburg, Russia
[3]Institute of Philosophy and Law, Ural Department of the Russian Academy of Sciences, 620144, Yekaterinburg, Russia

Abstract. This article considers the most important conditions for the creationof energy-saving competence of students – undergraduate training. The actualityof the problem is caused by low energy efficiency of industrial production, significantly reduces the competitiveness of the domestic economy and, consequently, the need to intensify activities on the rational use of energy resources. The definition of energy-saving expertise, marked its structural components (value-motivational, cognitive activity-reflexive). Stages of creationof energy-saving expertise in system of continuous vocational training: diagnosis, motivation, learning and reflective-evaluative. The possibilities of educational technology and psycho designed to work with the psychological barriers and resistance to the process of regulation of energy consumption, the production facilities at its energy efficiency and energy saving. In particular, it is recommended to use active learning methods (discussion, games, analysis of problem situations, etc.), Information and communication technologies for the implementation of educational and research projects in the field of energy conservation, the development and implementation of interactive multimedia learning environments.

1 Introduction to the problem

In today's post-industrial society it becomes a problem energy-saving the survival of all humanity. This is evidenced by the world practice of rationing energy saving: UN documents, legislative measures of the European Union, USA, Japan, and the Federal Law of Russian Federation No. 261-FZ "About energy-saving..." (2009) and the federal government program of Russian Federation "Energy Efficiency and Energy Development" (2013).

The actualityof the problem of energy-saving is due to several factors. Energy efficiency can slow down the rate of consumption of energy resources, reduce the impact of production and consumption processes on the environment, preserve the stability

[a] Corresponding author: stepanova.step@mail.ru

of the biosphere. Energy efficiency in every sphere is essentially reduced to reduce the useless loss of energy due to its rational use.

The energy intensity of industrial production and social services in Russia is several times higher than global indicators making our products uncompetitive, and life is unacceptable energy wastefulness. Russian manufacturer spends on production products on average three times more energy than the average world country.

The main legislative measures remain unfulfilled. One of the key reasons - the underestimation of the importance and timeliness of formulation system of compulsory and voluntary energy management in enterprises, organizations and every individual. Improving the energy efficiency of natural resources necessitates the development of public cultural bases of energy resources consumption. It is necessary to overcome stereotype waste all and in all. Hence the need to create new ways of thinking power.

Reasonable power consumption can be a manifestation of a particular kind of competence - energy-saving expertise. Since the energy-saving activities are included in the different activities, it can be concluded that the relevant competence refers to the universal or key competences that go beyond the specific subject areas, and aimed at solving urgent social, economic and environmental problems. With regard to different types of professional activities, this expertise can be considered and as a base [1].

The extend of a problem

Analysis of the works of foreign authors shows that energy-saving is a subject of scientific interest, but is considered in the context of the internal resources of the individual. Most of the works devoted to the study of issues related to the way in which to maintain or preserve their personal resources in modern conditions. So, C.N. Macrae, A.B. Milne, C.V. Bodenhausen experimentally confirmed that an important role for the energy saving of the person playing the so-called stereotypes existence [2]. The transition from the preservation of internal resources to realize the need to maintain the external (natural) resources, as well as to the formation of this skill is not a subject of intense scientific interest.

In Russian literature the concept of energy-saving competence gained some meaningful representation [3, 4]. As a rule, this competence structure is analyzed through the prism of the activity of specialists mainly technical profile. And viewed in the context of energy education.

In this study, under the energy-saving awareness of competence refers to a person with respect to energy-saving activities in which it implements competency integrated in the professional qualifications and experience.

A necessary condition for the formation of energy-saving competence is the development of personal qualities necessary for the implementation of various types of practices from the perspective of a reasonable power consumption. The composition of these properties reveals the following energy-saving components of competence:

- Value-motivational component is aimed at energy-saving, which in turn includes a plurality of motives and values hat define the social and personal awareness of the need for rational use of energy resources;
- Cognitive component involves knowledge of the theoretical foundations of energy and resources, knowledge in the field of innovations aimed at improving energy efficiency;
- The activity component is represented by a system of skills and abilities to apply energy-ecological knowledge in various activities;
- Reflective component is cross-cutting and is implemented in an adequate self-esteem of man's relationship to the actual environmental problems of our time, the assessment

of cash knowledge and understanding of their own willingness to carry out professional activities in accordance with the principles of rational energy use.

Thus, the training of qualified personnel with "thinking energy-saver" and capable of timely, independent and creative adoption of energy-efficient solutions can not be based on the preferential development of theoretical knowledge. Formation of the corresponding competence requires a complex implementation of training and educational, classroom and extracurricular educational tools within the professional and educational space saving.

2 The model of formation of energy-saving competence

Structural-informative model of energy-efficient competence based on forecasting possible result and includes four interconnected blocks (Figure 1).

Task befor forecast unit determines the orientation of the development of follow-up and contains the requirements of employers to save energy; information on the receipt, transmission and use of sources of energy contained in the federal state educational standards and educational programs. The tasks are updated towards the goal. The components of energy-saving competencies characterize the logic of the educational process.

The implementation of the operational unit promotes activities to implement energy-saving activities in all areas of human life through the integration of teaching, research and socio-cultural activities that determines the meaning of the formation of energy-saving expertise.

Reflective unit includes monitoring the formation of energy-saving competence by continuous or repetitive data gathering, representing a combination of certain key indicators of energy savings. The effectiveness of energy-saving characteristic of formation of competence is the educational and professional readiness of students to implement a methodology of energy-saving activities in all areas of modern life.

The proposed model became the basis for the determination of ongoing formation of energy-saving competence of students of secondary and vocational schools, the selection of interactive forms and methods of training, development of diagnostic tools for monitoring the process and the result of the formation of energy-ecological competence.

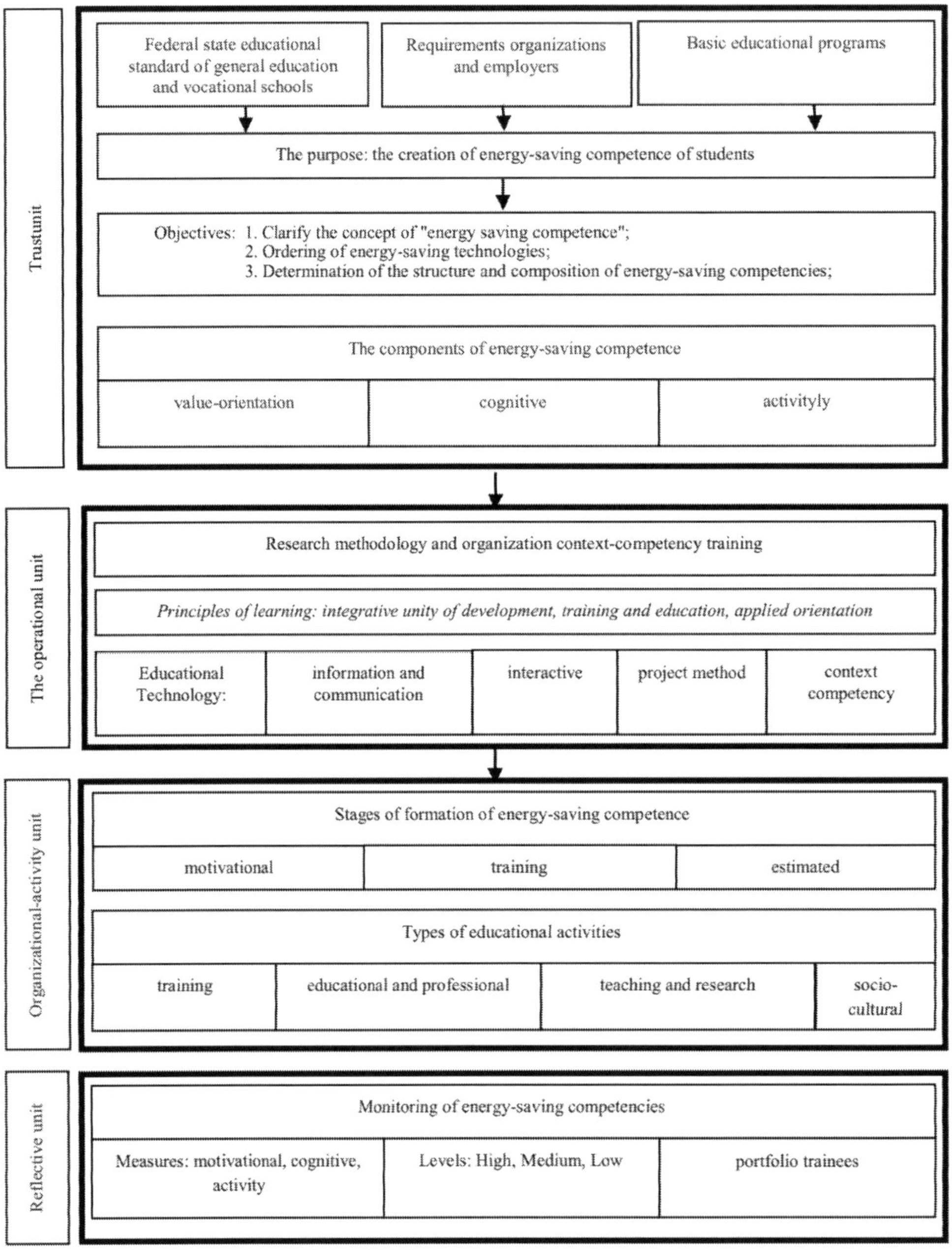

Fig. 1. Structural and informative model of energy saving competence of students in the (secondary), vocational school.

3 Educational technologies of formation of energy-saving competence

In the process of putting into practice the formation of energy-saving model of competence must take into account the nature of the activities of the future expert, highlight the content of the planned subjects invariant and variable component reflecting the specifics of the area of expertise. The important place is occupied by the use of educational technologies. In our opinion, for the formation of energy-saving expertise in vocational education are the most effective personality-oriented technologies used within the context of learning.

Contextual learning is a successful attempt resolution of the basic contradiction of vocational education lies in the fact that mastery of professional activities should be provided as part of educational activity, qualitatively different in content, forms, methods, tools and processes [5]. As a means of implementing the theoretical approach in contextual training is offered in full use of active learning methods, namely:

- Discussion techniques that can be implemented in the form of a dialogue, participants or groups of participants, Socratic conversation, group discussion or "round table", "brainstorming", seminar-debate, case studies, etc.;
- Play methods and forms of training can significantly improve the interest of students, to strengthen their attention, deeply understand the essence of the phenomena being studied, and submit yourself as a corresponding future professional activity;
- Project method: depending on the content of disciplines, within which will be carried out the project activities, the theme of research may involve the use of non-traditional (alternative) sources of energy, the introduction of automation devices, controllers, ensuring the optimization of technological processes and lead to a decrease in energy consumption, carrying out energy audit, the creation of modern systems of accounting and control of energy consumption, the development and support of new technologies aimed at improving energy efficiency, etc.;
- Constructive methods of learning and teaching pilot situational basis. Analysis of specific situations (case study) allows us to apply theoretical knowledge to solve practical problems in conditions as close to reality;
- The use of interactive electronic content: the advantage of multimedia learning environments is the ability to simultaneously reproduce the text on the computer screen, video and audio in some collection-related and logically subordinate to a certain didactic concept.

It should also be noted that the formation of energy-saving competence of students requires energy culture among teachers of vocational training. It is essential that teachers have broad erudition, freely oriented in problematic issues of energy, environment, legal and regulatory framework of energy, energy conservation in the home, in the workplace, in public services. This problem can be solved by improving the system of retraining and advanced training.

4 Conclusions and recommendations

Formation of energy-saving competence requires compliance with the principle of ecological conformity educational environment, a key component of which, in our opinion, should be the energy management.

Development of energy management, in addition to the optimization of the system (to achieve without major financial losses significant energy savings) associated with involvement in the energy management process of all participants in the educational process. The first step to understanding the objectives and principles of energy efficiency policy is transparency of energy consumption, open to students and teachers of educational institutions, the participation of students of technical training areas to monitor power consumption.

Organization of the system, a comprehensive work on the formation of energy-saving competence of the future teachers of vocational school promotes environmental awareness and implementation of state policy in the field of energy conservation and efficiency.

References

1. E.F. Zeer, E.E. Symanyuk, *Competence approach as the factor of innovative education* (2011)
2. C.N. Macrae, A.B. Milne, C.V. Bodenhausen, *Stereotypes as energy-saving devices: A peek inside the cognitive toolbox* (1994)
3. S.A. Rakutko, *Energy education: competence approach* (2010)
4. Y.I. Rakhimov, *Pedagogical technology of formation of professional competence of students of technical colleges energy* (2013)
5. A.A. Verbitsky, *Contextual learning: the concept and content* (2009)

E3S Web of Conferences **6**, 03008 (2016) DOI: 10.1051/e3sconf/20160603008
ICSC 2016

Energy efficient way of processing waste of milk production

Julija Panfilova[1a], *Maria* Ivantsova[1], and *Irina* Selezneva[1]

[1] Ural Federal University named after the first President of Russia B.N. Yeltsin, 620002, Yekaterinburg, Russia

Abstract. The research aim is the search of practical application of biotechnological waste processing of dairy production plants. The result of the study is the development of a process with biogas production by using anaerobic Biomar reactor. Scale of laboratory installation has been conducted based on laboratory studies. Moreover, principal technological scheme of production has been composed, and the appropriate material and technical and economic calculations have been carried out. In addition, major production facilities have been picked up. We found that using the produced biogas as a fuel for the boiler system allows you to reduce the natural gas consumption and reduce the cost of the recycling process at the dairy industry plants.

Whey is the main waste of dairy industry. Whey contains high concentrations of organic pollutants (fat, protein and lactose), inorganic compounds, synthetic surfactants and bacteria. Their content can be greater than the maximum permissible value.

In this work the object of laboratory tests was "Danone" plants waste water which contains cheese whey. The daily volume of water is 9.9 m^3. Cheese whey is a kind of whey which is colored liquid. The color varies from pale yellow to green. The chemical composition of cheese whey is shown in Table 1.

Table 1. The chemical composition of cheese whey

Solids in %	4,2–7,4
including:	
lactose (carbohydrates)	3.5
protein	0.8–1.0
milk fat	0.05–0.4
minerals	0.5–0.8
acidity, °T	50–85
density, kg/m^3	1019–1026
COC, mg/l	57100

The whey composition also includes vitamins E, C, B, and biotin (B7, H, coenzyme Q) and choline (Vitamin B4). Whey contains valuable mineral salts of phosphorus, magnesium and calcium.

[a] Corresponding author: Yusyaaaa@rambler.ru

Wastewater treatment methods are divided into mechanical, physical, chemical and biological ones. Biological methods of wastewater treatment are based on microorganisms that mineralize dissolved organic compounds, which are a power source for microorganisms.

Anaerobic decomposition of organic substances is carried out by the microorganisms' community that makes up the food chain of primary and secondary anaerobes. In contrast to the trophic chains of microorganisms in aerobic processes, where the relationships between groups of organisms are mainly characterized by relations "predator – prey", methane fermentation for trophic systems is characterized by the use of some groups of bacteria exchange products by the other ones.

Anaerobic methane fermentation involves four steps:

1. Step of enzymatic hydrolysis of undissolved complex organic compounds to formation of simpler dissolved substances;
2. Step of acid formation with the release of short chain volatile fatty acids (VFA), amino acids, alcohols and hydrogen and carbon dioxide (acidgenic step);
3. Acetogenic step of conversion of VFA, alcohols and amino acids into acetic acid, dissociating into hydrogen cation and acetate anion;
4. Methanogenic stage which is the formation of methane from acetic acid, as a result of the reduction reaction of carbon dioxide by hydrogen.

Five groups of bacteria involve in the process of anaerobic fermentation. The first group includes fermentative bacteria, represented mainly by Bacillus, Micrococcus, Pseudomonas and Clostridium, realizing the step of enzymatic hydrolysis and acid formation. Almost all of these bacteria are referred to growing facultative anaerobes with an optimum pH 6.5–7.6. Bacteria isolate biological catalysts called exoenzymes, which help realize the hydrolysis and transfer of solid insoluble compounds into the dissolved state. The rate of hydrolysis depends on the nature of the organic substances and of its conditions: namely it is necessary to provide sufficient enzyme amounts, create conditions for their contact with an organic substrate, support temperature and optimum pH.

Acidgenic step is carried out by the above heterogeneous groups of microorganisms for which carbon having passed into the solution of simple substances is a source of power. 20% of the original substances of original waste are converted into acetic acid, 15% are converted into propionic acid, and 65% are converted into other intermediates. Step of acid formation does not limit the subsequent steps of fermentation, because bacteria are unpretentious and grow at a high speed. However, actively proceeding steps hydrolysis and acid formation can lead to accumulation of volatile acid and decrease the pH, which is the direct cause of inhibiting the bacteria growth of subsequent process steps [1].

Acetogenic step is carried out by two groups of acetogenic bacteria. The first group forms acetate with hydrogen evolution (acetogens forming hydrogen are the second group) from the soluble product of the preceding step of acid formation:

$$CH_3CH_2COOH + 2H_2O \rightarrow CH_3COOH + CO_2 + 3H_2 \quad (1)$$

$$CH_3CH_2CH_2COOH + 2H_2O \rightarrow 2CH_3COOH + 2H_2. \quad (2)$$

The second group of acetogenic bacteria leads to the formation of acetic acid by the use of hydrogen for CO2 recovery (acetogens using hydrogen are the third group):

$$4H_2 + 2CO_2 \rightarrow CH_3COOH + 2H_2O. \quad (3)$$

At the fourth methanogenic step, methane bacteria produce methane in two ways namely by splitting acetate (a fifth group of bacteria) and reduction of carbon dioxide by hydrogen (a fourth group of bacteria).

$$CH_3COOH \rightarrow CH_4 + CO_2 \ (72\% \ CH_4) \quad (4)$$

$$CO_2 + H_2 \rightarrow CH_4 + H_2O \ (28\% \ CH_4). \quad (5)$$

5 major groups of methane bacteria namely Methanococcus, Methanobacterium, Methanospirillum, Methanotrix and Methanosarcina can be involved in the process. Methane bacteria are strict anaerobes.

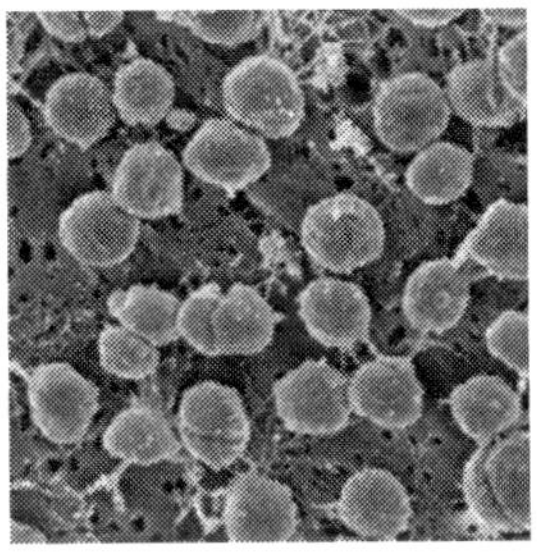

Fig. 1. Methanococcus jannischii.

Fig. 2. Methanosarcina archaea.

They are very sensitive to the presence of dissolved oxygen and nitrates in the medium. Optimum pH for them is 7.0–7.5. Carbon sources for methane bacteria are acetate ion and carbon dioxide. The energy source is hydrogen and the main source of nitrogen is ammonia. The main sulfur source is sulphides, although cysteine and sulfates can be the source too. Methanogens also need various trace elements (K, Na, Mg, Co, Cu, B, Zn, Mo).

Methane fermentation should always consider not individual groups of bacteria, but the entire community as a whole. The efficiency of fermentation process in such a community of bacteria does not only depend on the activity of the organisms involved in the reaction, but on the vital functions of bacteria that consume the products of this reaction. The accumulation of metabolic products of one of the process steps leads to the inhibition of others. Bacteria at different steps have their morphological and physiological characteristics, expressed in different growth rates, sensitivity to pH and O2 and others. All of these factors play a major role in creating a well-balanced and well-working system [2].

BIOMAR anaerobic treatment facilities are encouraged to use for organizing the biological wastewater treatment process in the dairy industry (Figure 1). Qualitative averaging of waste water is due to a specially calculated volume in the mixer. Sharp fluctuations in pH are naturally smoothed out, uneven receipt for cleaning drains is prevented. Another important function of the averager is the biological pre-oxidation of organic substances in the waste water. Duration of the water presence in the neutralizer for pre-oxidation was calculated based on the degree of contamination.

DOI: 10.1051/e3sconf/20160603008

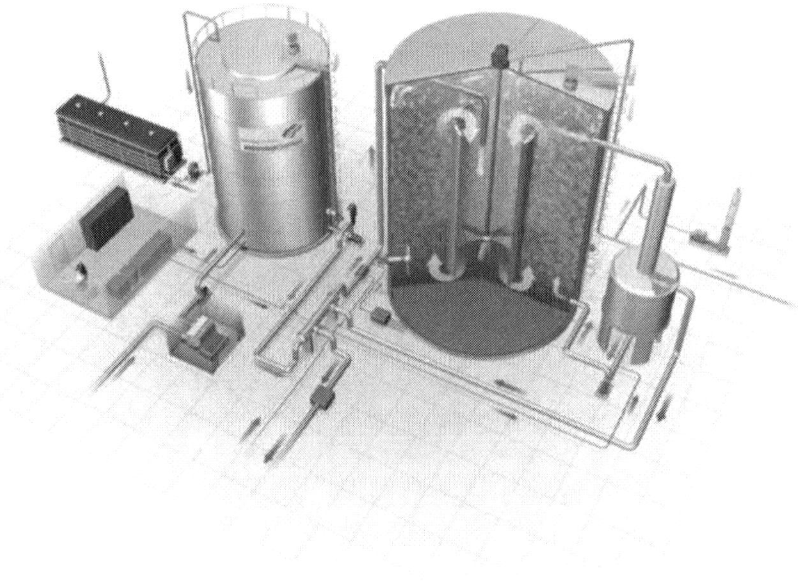

Fig. 3. Biomar ALB: anaerobic method using upflow reactor.

Before the wastewater gets into the anaerobic Biomar reactor, its optimal parameters (pH = 7.2–7.4 and T = 37 °C) are set automatically [1]. It should be noted that the use of the Biomar reactor will increase the performance of treatment facilities, improve the quality of waste water treatment and reduce the amount of excess sludge.

Another important advantage of Biomar reactor is a little need for production facilities, which allows the reconstruction of the existing plants.

Sediments are produced in the process of wastewater treatment. These sediments are subject to decontamination, disinfection, dehydration and drying. In addition, subsequent disposal of sediments is possible. Sediments contain significant amounts of fertilizing macro- and micronutrients, especially nitrogen, phosphorus and potassium. Moreover, the sediment includes zinc, manganese, copper, molybdenum and others. Sediments are usually utilized after their dehydration.

Dehydration of digested sediment is carried out with the help of mechanical apparatus namely vacuum filters, centrifuges, filter presses of different types, and drying in natural conditions in the sludge beds [3]. Digested sediment has high moisture content (95–98%), which makes it difficult to use it in agriculture. Humidity is the main factor determining the amount of sediment. Therefore, the main aim of sediment treatment is to reduce its volume due to water separation and production of a transportable product.

To solve these problems we have proposed a system of buildings and equipment for the anaerobic treatment of waste water and sediment treatment, which includes:

1) methane tank for anaerobic digestion;
2) the absorber to trap hydrogen sulfide;
3) gas tank for collecting the evolved biogas;
4) centrifuge for dehydration of digested sediment;
5) the dryer for driving sediment into a form suitable for transportation and later use.

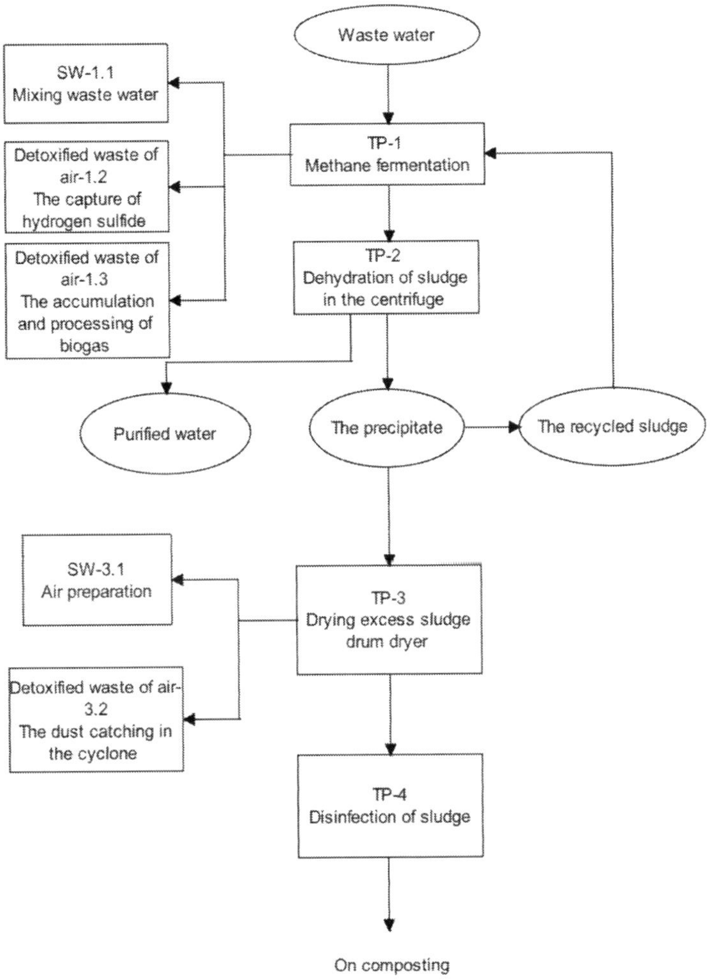

Fig. 4. Technological scheme of the process of wastewater treatment and disinfection of sediment

Along with the biogas, hydrogen sulfide is released during the methane fermentation, which is caught by monoethanolamine solution in the absorber of plate type. The cleaned gas from the absorber is collected in the gas holder. The saturated solution from the absorber after passing a heat transfer apparatus is sent to a desorber (distillation column) of packed or plate-type, where regeneration of monoethanolamine occurs.

$$HOCH_2CH_2NH_2 + H_2S \rightleftharpoons (HOCH_2CH_2NH_3)_2S$$

$$(HOCH_2CH_2NH_3)_2S + H_2S \rightleftharpoons 2HOCH_2CH_2NH_3HS$$

$$HOCH_2CH_2NH_2 + CO_2 + H_2O \rightleftharpoons (HOCH_2CH_2NH_3)HCO_3$$

The product of designed technological process is purified waste water. The main indicators of treated waste water quality are shown in Table 2.

Table 2. Indicators of treated wastewater

Display	Content in the purified waste water
Phosphates mg/l	3,5

Ammonium nitrogen, mg/l	1,5
pH	7,3

A byproduct of milk production wastewater is biogas that is reasonably expected to use as a fuel. Natural gas consumption will be decreased by using the generated biogas (we calculated that the daily volume of the methane formed is 70 m^3). The cost of natural gas consumed by plants is 4,011 rub/m^3. Calorific values of natural gas and methane respectively are 39800 kJ/mol and 40,000 kJ/mol. Technical and economic calculations have shown that annual savings from the use of biogas methane as a fuel instead of natural gas will amount to 100 thousand Rub.

It is possible to achieve the values of the monitored parameters, which do not harm the environment and will help to send treated water to the urban waste treatment plants as a result of the projected system construction for cleaning waste of the milk production.

Thus, it can be concluded that the biological anaerobic technology of wastewater treatment has lower operating costs compared to other methods. Moreover, the use of biogas would reduce the steam or electricity consumption by 10–15%.

References

1. L.I. Gunther, L.L. Goldfarb, *Methane-tanks* (1991)
2. L. Gunther, *Urban sewerage*: Collection of scientific papers, **VI**, 158–170 (1961)
3. M. Chebotaeva, *Wastewater treatment: trans. from germ* (2013)

E3S Web of Conferences 6, 03009 (2016) DOI: 10.1051/e3sconf/20160603009
ICSC 2016

ICSC – Policy for energy saving and increase of efficiency in Russia in the spheres of construction, housing and community amenities

Andrey Boyarinov[1a], *Iuliia* Plastinina[1], and *Nataliia* Dukmasova[1]

[1]Ural Federal University, Department of Environmental Economics, 620002, Yekaterinburg, Russia

Abstract. Russia's GDP energy intensity today is approximately 2.5–3.5 times higher than that of economically developed countries. To increase its economic competitive ability, Russia needs to achieve energy efficiency in different spheres, including construction, housing and community amenities. Close examination of implemented management measures and world experience revealed that in order to achieve a further energy efficiency increase Russia needs to boost economic interest of the participants concerned and to form effective mechanisms of economic management, and this should be done along with improvement of administrative governance. The paper focuses on the barriers that hamper economic motivation and provides recommendations for energy efficiency increase. Even partial implementation of suggested measures, in our opinion, will increase energy efficiency in the spheres of construction and housing accommodation, which is illustrated through the example of a residential building located in Yekaterinburg (Russia, Middle Urals).

1 Energy saving governance in Russia in 1998–2016

Energy saving is one of the rational ways to achieve economic management efficiency on an individual country scale, as well as stable development on a global scale. This can be explained by the fact that energy saving costs are 2–6 times lower than the costs of generation of the same amount of energy due to commissioning of new capacities, and, as this takes place, original energy resource is being saved. Today it is safe to say that the efforts made by foreign developed countries several years ago and aimed at improvement of energy saving mechanisms have become an advanced trend, which has been duly appreciated by Russia among other countries.

The first Russian state-run energy saving program for 1998–2005 revealed imperfection of governance and legislative framework in this sphere marked by general production decline and growth of energy consumption during that period of time. By the end of this decade, in order to achieve energy saving efficiency at par with economically developed countries, Russia must reduce economic energy intensity by at least 40% as of the 2010 level. This is required by the Decree of the President of Russia No. 889 (2008),

[a] Corresponding author: au.boyarinov@net-ustu.ru

as well as by the second state-run program known as *Energy Saving and Energy Efficiency Increase for the period of up to 2020* (state programme of the Russian Federation «Energy-saving and energy efficiency increasing for the period up to 2020», 2010) and its latest version (state programme of the Russian Federation «Energy efficiency and power industry development», 2014). However, energy efficiency of the Russian economy in the period from 2010 to 2014 has only seen a 13.3% increase, and even less than that, i.e. 5.61% increase, if we compare the year 2014 with the pre-crisis year 2008 (see Figure 1).

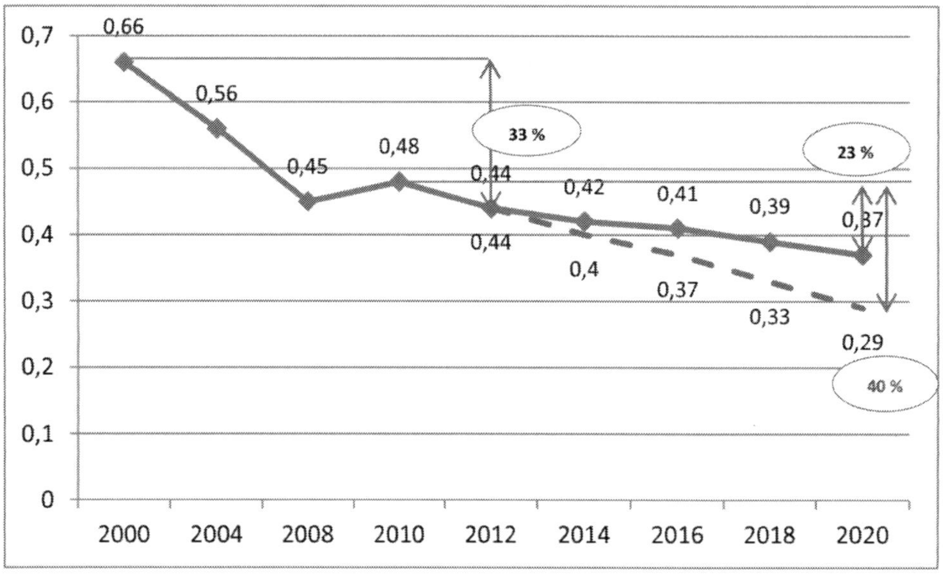

Fig. 1. Energy intensity changing per unit of GDP in Russia from 2000 to 2020, tons of equivalent fuel per $ 1,000.

Such rates are obviously insufficient to achieve the stated objectives. This is what experts define as inertial path of development, which potentially can reduce GDP energy intensity of Russia by only 23% instead of desired 40% (up to 0.37 TCE/$1,000 instead of 0.29). Factor analysis of energy intensity decrease in 2000–2012 (0.66 and 0.44 TCE/$1,000 respectively; 33%) carried out by the Russian Ministry of Energy explains the reasons for such inertia. As this analysis shows, it was not the progress in science and technology that became the main factor influencing the decrease rates, but the changes occurred in the structure of economy at the level of sectors and sub-sectors (Figure 2).

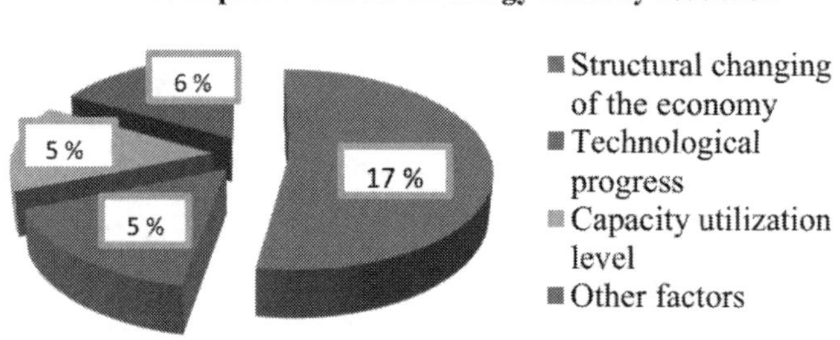

Fig. 2. Key factors of the energy intensity reduction per unit of GDP in Russia from 2000 to 2012, %.

Advanced energy saving technologies may enhance the interest of the participants concerned. Under market economy conditions, this means, first of all, economic interest. With this in mind, let us develop a potential action plan for energy efficiency in an industry that is vital for the economy of the country, such as construction and in an energy-intensive industry such as housing and community amenities (Figure 3).

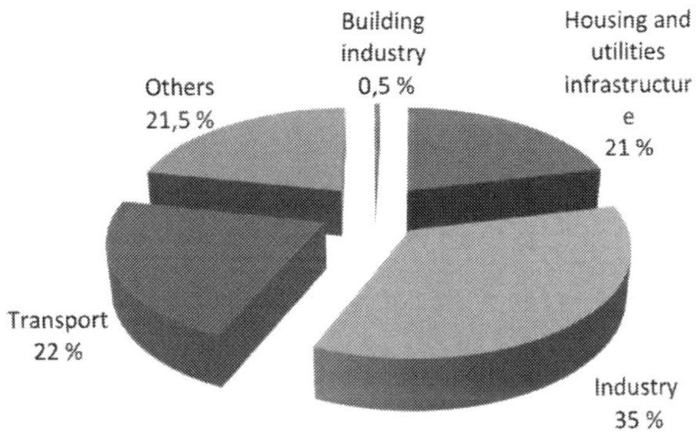

Fig. 3. Structure of the Energy balance of the Russian Federation, %.

First of all, let us describe peculiarities of construction, housing and community amenities in Russia. These industries in Russia have historically been subject to a strong state control. Although they develop in conformity with the laws of market economy and are profit-motivated, an administrative component is also very important for effective governance. One more peculiarity of the above mentioned industries is a high percentage of non-energy efficient buildings. Finally, a key aspect, specific not only to Russia, is that a real estate item undergoes change of ownership during its life: the ownership is transferred from the developer and/or investor (construction) to the owner/apartment complex manager (represented by municipal services in the majority of cases).

In other words, till the onset of market self-regulation, energy saving increase in the spheres of construction, housing and community amenities needs proper guidance with simultaneous provision of economic incentives for all the participants.

2 Analysis of energy saving administrative governance in the spheres of construction, housing and community amenities

At the moment, in order to ensure state control in the spheres of construction, housing and community amenities in Russia, administrative and economic instruments are applied (such division is rather conventional). Besides the development of main legal acts and energy saving planning (state programms «Energy...» and Federal Law No. 261 «On energy-saving and energy efficiency increasing and on making changes to the certain regulations of the Russian Federation», 2009), the following has been implemented:

a) Requirements for energy efficiency of buildings, newly built or already in use, have been defined.

b) Mandatory energy audit and energy certificates have been introduced for the buildings under construction and the buildings undergoing modernization and repairs.

c) Conformity to energy saving requirements is supervised by respective government agencies.

d) Administrative violations in the sphere of energy saving have been defined.

e) Budgetary support for federal and regional Energy Saving Programs has been implemented.

f) Reform of the sphere of housing and community amenities has been completed; instrument metering of the resources consumed has been introduced (mandatory for water and electric power so far).

However, these measures are not sufficient. For example, to achieve the desired level of energy resources consumption, a step-by-step transition from basic values of energy efficiency to the higher ones is envisaged for real estate items. In accordance with *Regulations of Energy Efficiency Requirements,* annual specific energy consumption values must be reduced at least once every five years starting from 2011 for buildings, constructions and structures: up to 15% with regard to base level till the end of 2015, starting from 2016 – up to 30%, and from 2020 – up to 40% (Decree of the Government of the Russian Federation No. 18 «On approval of the Regulation of establishing energy efficiency requirements for buildings, constructions, structures and the requirements to the rules determining the class of energy efficiency of apartment buildings», 2011). So, we see the policy of gradual energy "retooling" of the real estate items, though so far it covers only buildings under construction and buildings undergoing reconstruction and repairs. At the same time Capital Repairs Program developed for the period from 2015 to 2020 will cover only 4.5% of total amount of constructed buildings (Federal Law No. 261, 2009). The owners of the real estate items exempt from renovation and repairs can increase their energy efficiency on a voluntary basis at their own expense. In other words, the current situation with existing buildings does not contribute to rapid growth of energy savings, as their total number is much higher than that of the buildings under construction.

For rapid expansion of energy saving technologies, special energy service contracts have been introduced abroad. However, this strategic way stipulated in federal law (Federal Law No. 261, 2009) is not being implemented due to lack of interest from both the customers and the executors. In view of the above, the Russian Ministry of Energy has decided to set up a 100% state-run energy service company with further possibility of setting up private-state companies. But in order to achieve this goal, we will need to travel the long path of introduction and approbation of novelties with due regard to peculiarities of the Russian economy, therefore no rapid results can be expected in the near future.

3 Economic governance of energy savings in the spheres of construction, housing and community amenities

Since administrative governance has not yielded desired results during the initial period of implementation of the second energy saving program (from 2010 to 2014; Figure 1), the emphasis must be made on economic governance mechanisms, which is envisaged by market economy.

The factors defining economic interest of different real estate participants are quite specific:

1) For the developer these are:

- Demand for energy efficient real estate property.
- Effective mechanisms of state incentives.
- Availability of technological novelties in the energy saving sphere and maturity of the market of their production.

2) For the buyer these are:

- Information awareness of advantages of this category of dwelling.
- Price affordability and free consumer choice.

3) For the real estate owner/apartment complex manager these are:

- State financial support for implementation of energy saving technologies.
- Developed service market of energy saving engineering support and its introduction.

Consumer demand for energy efficient real estate is a basic incentive for the developers. At the same time it is driven by economic interest of future real estate owners. However, its growth in Russia faces a serious obstacle.

Low rates of demand generation are attributed to relatively low tariffs for consumed resources, as compared to Europe. For example, gas tariffs in Russia are approximately 10 times lower than European ones, water tariffs are from 2 to 10 times lower. Only electric energy tariffs are more or less the same. However, an actual housing and community amenities affordability index is represented not by the not by the tariff rates, but by the total amount of utility charges expressed through the percentage which the utility bills represent in the family budget. This index (16% of an average salary) places Russia slightly on top of the better-off Europe (up to 10–13%), though the tariffs indicate that it should be opposite [1]. This happens due to the fact that despite relatively low tariffs, utility norms in Russia assume increased consumption of the resources [2]. For example, in Yekaterinburg the regulatory standard (Standards of consumption of utilities approved by the Resolution REC Sverdlovsk region No. 131-PC, No. 132-PC, 2012) stipulates per capita consumption of hot and cold water as about 300 l per 24 h (for the apartments with a bath) in equal proportions, while thermal energy for heating and hot water supply as 1.3 Gcal per month.

We believe that normalization of consumed volumes of resources and their tariffs applied with due regard to a multiplying factor for overconsumption, as well as improvement of housing and community amenities must yield substantial results in demand generation for energy efficient apartments and resource savings due to development of explicit economic motivation.

Besides tariff and resource normalization, financial incentive of energy saving measures implementation might be quite effective as well. At the moment financial incentive is especially appropriate for owners of the apartments older than 1998, as some energy saving measures have high implementation costs. For example, heating modernization (installation of thermal energy meters and heating controllers) equals half the monthly income of the owner and requires even greater expenses in old buildings.

Germany managed to cope with this problem thanks to an innovative finance system developed by the German state bank for development *Kreditanstalt für Wiederaufbau (KfW) Förderbank (KfW)* within the framework of *Residential Stock Modernization Program* [3]. The aim of this Program is to render assistance to the owners with regard to initial expenses, including replacement of heating systems, windows and thermal insulation of external walls. As a rule, *KfW* offers beneficial loan conditions for its client: low and fixed interest rates, longer terms, possibility of redemption without any penalties, high maximum loan limit and possibility of its combination with other financial loans. Any residential project meeting energy saving requirements can seek a *KfW* low interest loan [4]. This finance system has proven successful in other EU countries as well.

Financial incentive is also important for the developer. In the situation of a vicious circle, when there is no supply due to lack of demand, and no demand due to lack of supply, government must intervene and act as a catalyst. Investment tax credit is a principal financial leverage anticipated for the developers in Russia. As stipulated in the Tax Code of Russia (para 1, Article 67), investment tax credit can be issued to any organization that constructs the items (including multi-dwelling buildings) of "the highest energy efficiency class A".

Both international and Russian experience prove that such a credit is beneficial, but Russia is facing two serious limitations today. Prior to issue a loan, the regulatory agencies verify financial position of a credit user, as well as its eventual organization expenses, which is a rather disturbing than well-established practice in our country. Besides, the highest class of energy efficiency (class A) means that specific thermal energy consumption for heating, ventilation and hot water supply of the building is 45%

lower than the base level (power inputs for 1 sq.m), which will definitely require greater investments in technologies and will increase final cost of the item (Order of the Ministry of Regional Development No. 161 "On approval of the Rules determining the class of energy efficiency of apartment buildings", 2011). Thus, receiving of the investment credit requires preliminary calculations of economic benefit which is rather doubtful in today's situation. However, we should take into account that such requirements are not more strict than those applied in Europe, though the transition towards them in the EU countries was gradual in the environment of a more developed market.

4 Recommendations for energy efficiency increase in the spheres of construction, housing and community amenities

To find a path to increase energy efficiency, we should avoid following a passive inertial way of development and actively influence the energy efficiency processes, implementing innovative products among other things (Figure 2). Based on the results of our studies, we suggest the following recommendations (a sample estimated amount of finance is specified in parentheses for each recommendation):

1. Adjustment of normative consumption rates and tariffs for energy resources in the housing sector. As for the tariffs, general utility charges should be taken into account. Thermal energy tariffs in Russia are substantially lower than in Europe. Low costs lead to lack of incentives for energy savings. However, sharp increase of tariffs may lead to popular dissatisfaction and financial problems in the economy of the country, particularly in the spheres of construction, as well as housing and community amenities. In view of the above, we suggest setting an annual adjustable pace for tariff growth for normative consumption with regard to predictive inflation rate, set forth by the Russian government. Overconsumption must be charged with application of differentiated tariffs with overconsumption penalties. This will allow achieving the immutability principle of legal framework, establish conditions for stable operation of the business structures and encourage them to implement energy saving measures.

2. Development of an effective credit system of energy saving measures for apartment owners. German experience adopted by the majority of European countries can be taken as a basis (described above). Loans will be issued by the State Bank for a justified amount at a nominal annual interest rate (say, 50,000 RUR for a family with subsequent annual inflation adjustment. At the initial stage this amount will be sufficient to lay thermal insulation, install meters and ventilation in the apartment). The state will compensate the difference between the market rate and credit rate as it does now, for example, for agricultural organizations (amount of finance equals to 300 bln RUR for 10 years).

3. Improvement of the system of financial incentives for construction of energy efficient buildings. To achieve this goal, the government must increase financial appeal and reduce the number of documents that the organizations seeking investment credits and constructing Class A buildings must present. When constructing small apartment complexes and individual houses, the main problem, as we believe, arises when they need to be connected to power grids (electrical supply networks, gas pipelines, heat plants). Since the organizations responsible for such connections are de facto state-controlled, the bureaucratic procedures must be simplified and connection charges must be canceled for small private real estate owners (villa owners, country house owners). "Losses" incurred by utility companies can be compensated in this case by means of application of increased tariffs for real estate owners. Thus, connection charges will be collected by installment via application of increased tariffs (amount of finance equals to 125 bln RUR per year).

4. Support of production and sales of energy efficient equipment and engineering systems at the country level via application of a tax deduction program; support of local

DOI: 10.1051/e3sconf/20160603009

manufacturers via a state procurement program for such equipment and systems; this program must also cover the purchase of such equipment for capital repairs of multi-dwelling buildings at the expense of regional repair funds. Introduction of such equipment through a wider application of PPP energy service contracts and contracts signed with regard to market competition (amount of finance equals to 2 bln RUR per year, partially at the expense of capital repair funds).

5. Launch of the information awareness campaign for population, developers and property management companies. Government must launch the Russian integrated portal which will provide official detailed information about capital repairs and technical and economic energy saving novelties. The sites of property management companies must present information about organizations responsible for sales and installation of meters, installation works, thermal insulation, etc. Availability of such information and free consumer choice will form demand for energy-efficient goods and lead to quality improvement of the services offered in the energy saving sphere (amount of finance equals to 2 mln RUR per year).

Implementation of these recommendations can be funded through the costs planned for Sub-program 1 (*Energy Saving and Energy Efficiency Increase*) of the state program *Energy Efficiency and Energy Development*.

We believe that even partial introduction of these measures will increase energy efficiency in the sphere of construction and housing accommodation. Let us prove this point, taking Yekaterinburg as an example.

5 Determination of energy efficiency for a standard accommodation building located in Yekaterinburg (Russia, Middle Urals)

Energy efficiency is characterized by such principal energy and economic indicators as energy savings by volume and value, capital investment and payback period. Required scope of energy saving measures is determined by physical characteristics and state of the building. According to the regional energy saving program (Sverdlovsk region's programme "The energy-saving and energy efficiency increasing for the period from 2010 to 2015 and target settings for the period up to 2020", 2010), energy efficiency increase in the housing sector in Sverdlovskaya Oblast is equal to at least **5.240 mln TCE or 32,927.87 mln RUR.**

As a rule, existing non-energy efficient housing accommodation needs several mutually reinforcing measures: additional thermal insulation, installation of thermal energy flow regulation system for heating, modernization of ventilation and window frames. Besides, in accordance with legal requirements, the apartments must be equipped with meters for all types of resources, including thermal energy (i.e. heating and hot water supply). We calculate energy efficiency for each measure separately.

As an example, pre-feasibility study of the energy efficient modernization of a typical 3-storey building which consists of 36 apartments was produced. The total building square is 3767.7 square meters; the facade area is 2511.8 square meters; square of heated floors is 1980 square meters; the total heated building volume is 1320 cubic meters.

Selection of heat-insulation materials for the building was founded on the comparative feasibility study of the most popular brands of heat-insulation materials (Rockwool®) and foam polystyrene, as well as a new product of the Russian market - the HDPE membrane heater (Tyvek®) (Feasibility study of energy-saving measures and energy efficiency increasing, Cheboksary, 2014). These feasibility studies are summarized in the Table 1.

DOI: 10.1051/e3sconf/20160603009

Capital costs for one apartment are determined by dividing the common utilities cost by the number of apartments.

Table 1. Technical and economic justification for the choice of heat-insulating material.

Parameter	Reference	Polystyrene foam	Rockwool	Tyvek
Annual thermal energy savings due to reduction in heat losses through cladding structures, Gcal	ΔQ	175.78	199.035	206.01
Energy savings at a thermal source for thermal energy production, kWh	ΔE	3,164.11	3,582.64	3,708.2
Fuel savings at a power-supply source, TCE	ΔB_e	0.727	0.823	0.852
Fuel savings due to reduction of thermal energy consumption, TCE	ΔB_q	38.45	43.389	45.06
Cumulative fuel savings, TCE	ΔB $(\Delta B_e + \Delta B_q)$	39.18	44.212	45.92
Capital investment into the measure, RUR	C_i	5,023,600	2,906,403.8	4,822,656

Cost of purchase and installation of the thermoregulators for heat consumption control, modern air-handling unit, window frames with triple glazed insulating unit (three frames per one apartment, one of which comes together with a balcony door), meters for heat flow in central heating (3 thermal energy meters for the same amount of heating radiators per apartment, in average) and water meters (two per apartment) was determined on the basis of market price for the services and most popular models of the equipment available on the market.

Energy saving due to installation of a thermal energy flow regulation system for central heating, modernization of ventilation system and window frames was calculated with regard to heat consumption norms (average monthly consumption is 0.024 Gcal/sq.m; 8-month heating season), tariff rate and design procedure of heat loss. Calculation of economic efficiency for the complex of mutually reinforcing energy saving measures is laid out in Table 2.

Table 2. Economic justification of energy efficient modernization of one apartment building in Yekaterinburg.

Measure/ Energy and economic indicator	Items for investment	Capital investment, RUR	Energy savings, Gcal / %	Energy savings, RUR	Payback period, years
Thermal insulation	Purchase of mineral wool insulator *Rockwool* for external walls, attic floor and basement (295.4 RUR / sq.m); installation (180 RUR / sq.m); daubing of a front (consumption of a stucco – 1,8 kg / sq.m; cost – 70–90 RUR / kg; basement – 180 RUR / sq.m) total area with cladding 3,767.7 sq.m; front area 2,511.8 sq.m.	3,767.7 • 295.4 + 3,767.7 • 180 + 2,511.8 • 1.8 • 80 + 2,511.8 • 300 / 36 = 1,112,978.58 + 678,186 + 361,699.2 + 753,540 / 36 = **80,733.44**	5.53 Gcal / 25 %	5.53 • 1,579.18 = **8,730.89**	9.25
Heat	Thermoregulator *Danfoss*	1,876.54 • 3	4.23 Gcal /	4.23 •	1.4

consumption control through heating systems	(valve and thermostat – 1,876.54 RUR); installation (about 1,200 RUR); 3 widgets / apartment; average monthly consumption is 0.024 Gcal / sq.m; tariff rate – 1,579.18 RUR / Gcal; area apartment – 42 sq.m	+1,200 • 3 = **9,229.62**	35 %	1,579.18 = **6,685.61**	
Modernization of ventilation system	Purchase of Ventilation installation with a recuperator Systemair VX (61,000 RUR); installation (10,000 RUR)	**71,000.00**	4.8 Gcal / 25 %	4.8 • 1,579.18 = **7,580.1**	2.5
Modernization of window frames	Purchase and installation of triple -glass pane (1,400 x 1,600 mm – 11,650 RUR; 700 x 2,200 mm – 8,300 RUR); strength of heat transmission of ordinary window (SNIP 23-02-2003) – 06 (sq.m • °C) / Wt; strength of heat transmission of triple-glass pane (with I-glass, material – plastic; Ar) – 0.90 (sq.m • °C) / Wt	11,650 • 3 + 8,300 = **43,250**	0.621 Gcal / 30 %	0.621 • 1,579.18 = **980.67**	14.7
Installation of meters	Purchase of thermal energy meters with the set of respective equipment (11,000 RUR); installation (4,000 RUR); 3 widgets; average monthly consumption is 0.024 Gcal / sq.m; tariff rate – 1,579.18 RUR / Gcal; area apartment – 42 sq.m	15 000 • 3 = **45,000**	– / 40 %	0.024 • 42 • 1,579.18 • 12 • 0,4 = **7,640.7**	5.9
	Purchase of water energy meters (600 RUR), 2 widgets; installation (800 RUR); tariff rate – 1,579.18 RUR / Gcal; average monthly consumption of a water is 5.61 cub.m / man or 0.583 Gcal; live 2 mans	600 • 2 + 800 • 2 = **2,800**	– / 75 %	0.583 • 2 • 1,579.18 • 12 • 0,75 = **16,628.8**	0.17
Total		**252,013.06**	**23.13 / -**	**48,246.77**	**5.22**

Thus, introduction of even a minimum required complex of energy saving measures, many of which can be implemented by the occupants at their own expense, will ensure 14% energy savings and significant cost savings. With energy efficient equipment available on the market, especially high quality domestic models, savings by volume and by value can be even more substantial.

6 Conclusion

Elaboration of recommendations aimed at energy efficiency increase in the spheres of construction, housing and community amenities in Russia has yielded the following conclusions.

Energy efficiency governance in these spheres must take into account peculiarities of the real estate market, i.e. its dependence on the administrative control and the presence of several key participants (such as the developer, investor, buyer, real estate owner (represented normally by a property management company)). Administrative control has proven to be inefficient and needs additional economic mechanisms. Principal barriers for the development of economic interest of the participants of the real estate market are as follows: low demand, insufficient information awareness of the key participants, insufficiently effective incentives or their total absence, bureaucracy of natural monopolies and regulatory agencies in the sphere of construction.

We recommend the following to tap energy savings:

1) Adjust tariffs for energy resources in the housing sector with due regard to general utility charges.

2) Develop an effective credit system of energy saving measures for real estate owners.

3) Improve the system of financial incentives for construction of energy efficient buildings.

4) Enable legislation in support of production and sales of energy efficient equipment and engineering systems.

5) Introduce such equipment through a wider application of energy service contracts signed with regard to market competition.

6) Launch an information awareness campaign for population, developers and property management companies.

Implementation of these suggestions will help achieve world level of energy savings due to introduction of innovative solutions, thus supporting and developing domestic economy.

References

1. D. Sergeeva, House and Yard, **15/05/2014** (2014).

2. A. Kolesnikov, *Energy-saving on the industrial plants and utilities: tutorial* (INFRA, Moscow, 2008)

3. International Energy Agency, *Progress with implementing energy efficiency policies in the G8* (OECD/IEA, 2009)

4. H. Schettler-Köhler, *EPBD Buildings Platform*, **P 73-18.04.2008**, 1–7 (2008)

E3S Web of Conferences **6**, 03010 (2016) DOI: 10.1051/e3sconf/20160603010
ICSC 2016

The greenhouse index of sustainable development for metallurgical processes of production in aspect of green power

Vladimir Lisienko[1], *Valery* Anufriev[2], *Dmitry* Berg[3], *Yuri* Chesnokov[1], and *Anna* Lapteva[1a]

[1]Ural Federal University, Institute of radio electronics and information technologies, Department Automatics, 620002, Yekaterinburg, Russia
[2]Ural Federal University, Higher School of Economics and management, Department Economy of production and power systems, 620002, Yekaterinburg, Russia
[3]Ural office of the Russian Academy of Sciences "Institute of industrial ecology", 620002, Yekaterinburg, Russia

Abstract. The accounting of greenhouse gases (GHG) according to plans of the Russian Federation becomes obligatory since 2016 for the enterprises with issue of GHG not less than 150 thousand tons of CO2, since 2017 – from 50 thousand tons of CO2. Introduction of a carbon tax (15 dollars/t of CO2) is planned. Voluntary inventory of GHG of the enterprises and territories which in the long term apply is carried out to be called clever. According to University of Cambridge (USA) about 45% depreciation of joint-stock portfolios in the world markets, the expected climate change connected with are expected. Parameter for an assessment of the clever city with the developed metallurgy and need of decrease in greenhouse emissions in the atmosphere – the indicator of a sustainable development considering emission of carbon dioxide, prime cost of steel and its power consumption is offered. Its values for tandems blast furnace (BF) + oxygen converter (OC), BF, Corex, Romelt, Midrex, Hyl-3 everyone with arc furnace (AF) are defined.

The modern "clever" cities are not only environmentally friendly cities, but also the cities in which it is developed industrial, including, metallurgical production. At the same time the problem of emission of the GHG exerting a great influence on an ecological situation is of great importance. The submitted report just is also devoted to the choice of technologies with the minimum emission of GHG.

The accounting of GHG according to plans of the Russian Federation becomes obligatory since 2016 for the enterprises with issue of GHG not less than 150 thousand tons of CO2, since 2017 from 50 thousand tons of CO2. Introduction of a carbon tax is planned, the preliminary cost of 1 ton of CO2 is designated in 15 dollars. Voluntary inventory of GHG of the enterprises and territories which in the long term can apply for recognition by clever is carried out. According to University of Cambridge (USA) about 45% depreciation of joint-stock portfolios in the world markets, the expected climate change connected with are expected.

[a] Corresponding author: annalapteva@mail.ru

Green technologies and sustainable development is the principles at which development becomes steadier, and resources are used more rationally. They assume decrease in negative impact on environment – is the main line of green technologies. This development is "proceeding" ("self-sufficient"), that is such which doesn't contradict further existence of mankind and its development in the former direction. In this plan it is known that the emission of GHG which is directly connected with power component, can make negative impact on a condition of climate. To UrFU the break-through technique of the accounting of GHG so far on the example of the enterprises the supplementing a technique of Intergovernmental Panel on Climate Change (IPCC) is mastered [1]. It is based on use of indicators of physical and chemical processes and gave more exact results when calculating GHG at the level of units (AF, OC, etc.) and the subsequent decrease in issue of CO2 at energy efficiency increase. Further this technique can be adapted for color metallurgy and chemical industry. Such approach is very important for transformation to the clever cities industrial and monotowns, such as Nizhny Tagil, Magnitogorsk, Lipetsk, Cherepovets, etc.

This work is also devoted to a problem of emission of GHG – carbon dioxide – when functioning one of the most power-intensive branches of the industry – ferrous metallurgy. There is a problem of formalization of definition of green technologies. For these purposes index of sustainable development – the indicators allowing to judge a state or change of an economic, social or ecological variable serve.

The index of sustainable development determined by a formula is offered:

$$K_E = \alpha_1 \cdot y_1 + \alpha_2 \cdot y_2 + \alpha_3 \cdot y_3 \quad (1)$$

where $\alpha_1 + \alpha_2 + \alpha_3 = 1$ – weight coefficients; y_1 – relative through emission of carbon dioxide; y_2 – relative prime cost of the product received in some processes; y_3 – relative through power consumption of this product.

Relative through emission of carbon dioxide is defined as

$$y_1 = I_i / I_{max} \quad (2)$$

where I_i – specific through emission of technological process or a carbon trace, kg on t of production; $I_{max} = 3000$ – the greatest possible through emission of carbon dioxide in the considered processes, kg/t of production.

Relative prime cost of a product is determined by a formula

$$y_2 = PC_i / PC_{max} \quad (3)$$

where PC_i – prime cost of a product in i process, ruble on t of product; $PC_{max} = 30{,}000$ – the maximum prime cost of a product, ruble on t of product.

Relative through power consumption of a product is determined by a formula

$$y_3 = TFN_i / TFN_{max} \quad (4)$$

where TFN_i – specific power consumption of a product in technological fuel numbers, kg of equivalent fuel on t of production; $TFN_{max} = 1{,}500$ kg of equivalent fuel/t of production – the maximum power consumption of production.

The offered index of sustainable development characterizes processes from the following points of view: emissions of greenhouse gas, expenses of work and materials, consumption of energy resources. Such indicator can be used for an assessment of various processes. In this article it is applied to an assessment of processes of production of cast iron and steel. In these processes the main formed greenhouse gas is carbon dioxide as other emitted gases (methane, carbon oxide, hydrogen) are a part of the secondary energy resources which are used in metallurgical processes as fuel, burning down to carbon dioxide and water.

The following metallurgical conversions are allocated for which index of sustainable development are calculated: BF, Corex, Romelt, Midrex, Hyl-3, OC, AF [1–4]. In processes of BF, Corex, Romelt cast iron is melted. In processes of Midrex, Hyl-3 sponge iron is formed. Cast iron or sponge iron are loaded into OC or AF in which steel is melted. The following tandems of technologies are analysed: BF + OC; BF + AF; Corex + AF; Romelt + AF; Midrex + AF; Hyl-3 + AF. At the exit of all combinations of repartitions for comparability the same product – crude steel was accepted.

Prime cost of crude steel is taken from [5]. Product cost in Russia in the present is characterized by considerable instability.

The technique and formulas for calculation of through emission of carbon dioxide are given in works [6–10].

The technique and formulas for calculation of TFN are taken from [11–14]. Calculations are carried out with the following values of coefficients: $\alpha_1 = 0.3$; $\alpha_2 = 0.5$; $\alpha_3 = 0.2$. The priority is given to prime cost. Results of calculations of values of the index of sustainable development are given in table 1.

Table 1. Ranging of processes on values of the index of sustainable development K_E.

Designation of a technological chain	Index K_E	Emission of carbon dioxide, kg /t of steel	Prime cost of steel, ruble/t of steel	TFN, kg of equivalent fuel/t of steel	Priority on		
					K_E	TFN	prime cost
AF on the scrap	0.534	1,024	20,649	441	1	1	7
Hyl-3 + AF	0.562	1,203	18,997	630	2	2	2
Midrex + AF	0.577	1,258	19,376	639	3	5	3
BF + AF	0.602	1,434	20,123	634	4	4	6
Corex + AF	0.703	1,904	16,380	632	5	3	1
Romelt + AF	0.770	2,036	19,982	911	6	6	5
BF + OC	0.733	2,182	19,893	917	7	7	4

In table 1 technological chains are located on increase in values of the index of sustainable development. As appears from calculations, the steel melted in AF from metal scrap has the minimum value of the index of sustainable development. As we see, AF on scrap, Hyl-3 + AF, Midrex + AF, tandem BF + AF at prevalence of scrap in furnace charge of AF enters in the five of the most priority processes on this indicator.

Thus, as appears from table 1, on values of the index of sustainable development first place is won by the processes which aren't connected directly with production of cast iron. Further processes in which in a varying degree in initial furnace charge cast iron is used follow.

Results of calculations with the following values of coefficients: $\alpha_1 = 0.3$; $\alpha_2 = 0.2$; $\alpha_3 = 0.5$ are given in table 2. The priority is given to through power consumption.

Results of calculations with the following values of coefficients: $\alpha_1 = 0.5$; $\alpha_2 = 0.2$; $\alpha_3 = 0.3$ are given in table 3. The priority is given to through emission of carbon dioxide.

Table 2. Ranging of processes on values of the index of sustainable development K_E for values of coefficients $\alpha_1 = 0.3$; $\alpha_2 = 0.2$; $\alpha_3 = 0.5$.

Designation of a technological chain	Index K_E	Priority according to K_E
AF on the scrap	0.460	1
Hyl-3 + AF	0.562	2
Midrex + AF	0.574	4
BF + AF	0.562	3
Corex + AF	0.721	5
Romelt + AF	0.813	6
BF + OC	0.809	7

Table 3. Ranging of processes on values of the index of sustainable development K_E for values of coefficients $\alpha_1 = 0.5$; $\alpha_2 = 0.2$; $\alpha_3 = 0.3$.

Designation of a technological chain	Index K_E	Priority according to K_E
AF on the scrap	0.440	1
Hyl-3 + AF	0.516	2
Midrex + AF	0.531	3
BF + AF	0.562	4
Corex + AF	0.693	5
Romelt + AF	0.766	6
BF + OC	0.771	7

Conclusions

Within green power for the industrial city with the developed metallurgy the index of sustainable development taking into account emissions of carbon dioxide, prime cost and power consumption of made steel is being devised. The technique of their definition is offered. Calculations for this technique revealed tandems of repartitions which are the most preferable from the point of view of emission of carbon dioxide and product cost.

The smallest values of these indicators are inherent in processes of production of steel in which cast iron isn't used.

The economic (cost) essence of power-ecological and power-greenhouse analyses consists in close interrelation of concepts of these analyses and of expenses of the user of nature on a payment for emissions by performing of technological processes.

Change of values of weight coefficients slightly change a picture on values of indexes of sustainable development.

The recommended processes provide considerable decrease in greenhouse emissions within the industrial city and green power.

References

1. Yu.N. Chesnokov, V.G. Lisienko, A.V. Lapteva, Steel, **8**, 74–77 (2011)
2. V.G. Voskoboynikov, V.A. Kudrin, A.M. Yakushev, *General metallurgy* (Moscow, 1998)
3. V.G. Voskoboynikov, V.A. Kudrin, A.M. Yakushev, *General metallurgy*: The textbook for higher education institutions (Moscow, 1985)
4. Yu.S. Yusfin, N.F. Pashkov, *Metallurgy of iron*: The textbook for higher education institutions (Moscow, 2007)
5. V.A. Romenets, *Process Romelt* (Moscow, 2005)
6. V.G. Lisienko, V.V. Krivorotov, D.B. Berg, Yu.N. Chesnokov, A.V. Lapteva, Bulletin UFU: "Series Economics", **6**, 18–24 (2015)
7. V.G. Lisienko, Yu.N. Chesnokov, A.V. Lapteva, Metallurgist, **7**, 40–45 (2011)
8. Yu.N. Chesnokov, V.G. Lisienko, A.V. Lapteva, *Collection of scientific works of the All-Russian scientific and practical conference "Engineering Ecology"*, 50–54 (2011)
9. Yu.N. Chesnokov, V.G. Lisienko, A.V. Lapteva, Metallurgist, **12**, 23–26 (2012)
10. V.G. Lisienko, Yu.N. Chesnokov, A.V. Lapteva, Metallurgist, **5**, 18–24 (2015)
11. V.G. Lisienko, N.V. Solovyova, O.G. Trofimova, *Alternative metallurgy: alloying problem, model estimates of efficiency* (Moscow, 2007)
12. V.G. Lisienko, S.E. Rozin, Ya.M. Schelokov, News of higher education institutions. Ferrous metallurgy, **2**, 108–112 (1987)
13. V.G. Lisienko, O.G. Druzhinina, V.A. Morozova, News of higher education institutions. Ferrous metallurgy, **9**, 61–65 (1999)
14. V.G. Lisienko et al., News of higher education institutions. Ferrous metallurgy, **7**, 69–72 (1998)

E3S Web of Conferences **6**, 03011 (2016) DOI: 10.1051/e3sconf/20160603011
ICSC 2016

Towards a sustainable development given the gradual conventional energy sources upgrade

Vladislav O. Samoylenko[1a], *Stanislav A.* Eroshenko[1], and *Andrew V.* Pazderin[1]

[1]Ural Federal University, Department of Automated Electrical Systems, 620002, Yekaterinburg, Russia

Abstract. In spite of the successful renewable energy sources (RES) development, conventional energy sources still play a predominant role in meeting power and energy balances. So, the sustainable development in a short-term and middle-term perspective to be implemented by a gradual conventional energy sources upgrade along with RES deployment. The paper gives an analytical review of the different environmental aspects of the outdated conventional large-scale generation replacement by the up-to-date distributed generation (DG). The main goal of the paper is the consideration of an optimal strategy regarding conventional generation reconstruction and replacement to achieve the sustainable development of cities and suburban areas.

1 Introduction

The ongoing implementation of the renewable energy sources is supposed to be one the most perspective ways to improve the energy efficiency as well as to ensure a sustainable economy development. Nowadays RES in many countries provide a share of energy covering the electrical load of cities and suburban areas [1]. Also in warm climate RES may provide a great share of necessary heat energy.

Nevertheless, conventional energy sources still play a predominant role in meeting power and energy balances [1]. In cold climate different technologies like combined heat and power (CHP) are essential. In some areas RES application is restricted by the severe climate conditions and economical factors. The perspectives of a nuclear energy cause a discussion [2], therefore it is not available in many countries and areas.

So, in short-term and middle-term perspectives the conventional energy sources are expected to be the main sources, although sufficiently and steadily decreasing their share in meeting power and energy balances. It leads to the necessity of keeping existing generation centres on the high level from the point of view of energy efficiency and environmental impact.

The problem at hand is how to choose the optimal strategy regarding the reconstruction and replacement of conventional generation to achieve the sustainable development of cities and suburban areas.

[a] Corresponding author: vvsamoylenko@yandex.ru

It should be noted that both kinds of the sources are presented by large-scale bulk generation, like units of heat power plants or large wind farms, as well as by the distributed generation, although the latter is more typical for the RES. The types differ by their ecological impact features, energy efficiency, power quality and reliability [3]. One of the possible approaches to improve the considered aspects is to replace centralized system generation by the distributed one.

The paper gives an analytical review of the different environmental aspects of the conventional large-scale generation gradual replacement by the distributed generation. The aspects include total energy efficiency and fuel utilization factor (FUF), pollutant emissions and wastes decrease, water utilization. Also the paper describes a number of generation fuel technologies and primary sources from the point of view of their advantages and disadvantages. It is shown that the approach of the decentralization and downsizing effect is more suitable for localizing the impact than for decreasing it. A case study on the basis of the industrial Sverdlovsk region of Russia is provided.

The analysis presented in the review is supposed to be useful for energy companies, regulatory authorities and science representatives to choose the optimal decisions regarding the upgrade of the conventional generation.

2 Reducing the environmental impact of conventional generation

The typical energy sector and power industry development plans corresponds a strategy of a sustainable development and include the estimation of significant impact of energy companies on the environment.

The key points of a sustainable development strategy, including energy efficiency and environmental impact aspects, like reducing the volume of pollutant emissions into the atmosphere and the amount of disposed waste, are stated in the corresponding Concepts.

The environmental performance of the energy sector to be improved primarily by means of reconstruction and modernization of power plants. A special attention is paid for perspective generation and changes in the existing generation structure.

Excluding the RES, in the energy sector all over the world the share of hydrocarbon fuel generation dominates. It is represented mostly by various gas-fired generation, including relatively powerful combined cycle gas turbines (CCGT) as well as small-scale distributed generation like gas turbines (GT) and gas piston units (GP). At the moment it is associated with a wide range of generating capacity available for gas, simple gas fuel infrastructure use and specific indicators of required land.

Compared to the existing generation like bulk conventional steam-powered carbon and gas generation, new power plants have distinguishing environmental behavior that is necessary to take into account when planning the development of energy sector [4].

The aspects of the environmental impact of different conventional generation are presented in tables 1 and 2.

The data from the tables provide the following conclusions on the environmental criteria of the compared generation development to be taken into account when implementing the strategy of the sustainable development.

1. Up-to-date and perspective generation have higher than the existing conventional generation efficiency and fuel utilization factor; that leads to the less fuel amount burned and corresponding level of emissions and waste level reduction. Also, the new types of generation have improved specific indicators of pollutant emissions.

2. Carbon generation as well as the generation using local kinds of fuel, like based on pulverized coal burning (PCB), has high level of sulfur oxides and dust emissions. This is due to the properties of a coal and depends on its ash content. Modern facilities equipped

with dry ash removal and flue gas desulfurization significantly reduce this disadvantage, but require much more additional land for the allocation compared to gas station.

3. Modern powerful CCGT and high-power gas turbines (HP GT) have high efficiency. However, the trend of power gas turbines design leads to an increase in temperature and pressure in the combustion chamber to enhance the efficiency. It results in increased levels of nitrogen oxides NO_x emissions and the necessity to neutralize them.

4. Typically, it was believed that the share of monoxide CO emissions from conventional power plants is negligible due to the possibility of a flexible air supply dosing and control of the combustion process. However, taking into account new technology generation and a decrease in the share of other emissions, CO emissions become weighty, so they to be considered when developing generation facilities.

5. Hydrocarbon fuel distributed generation has advantages in comparison with the *existing conventional generation* by its environmental performance.

Table 1. The environmental impact of different conventional generation types

Name	Formula	The main source	The negative impact	Cumulative	Status
Gases and sprays					
Sulfur dioxide	SO_2	Coal and local fuels generation	Acid rains, corrosion, damage to plants	No	Local
Nitrogen oxides	NO, NO_2	All the high-efficiency generation	Carcinogens, corrosion, ozone layer destruction	No	Local
Carbon dioxide	CO_2	All the generation	The greenhouse effect	Yes	Global
Carbon oxide	CO	All the high-efficiency generation	Poison	No	Local
Solid particles	Dust	Coal and local fuels generation	Diseases	Partially	Local
Metals	Fe, Pb, Ni, Zn, Cu, Mn, Cr, …	Piston engines DG using oils	Poisons, carcinogens, mutagens	Yes	Local
Solid					
Wastes	Ash disposal area	Coal and local fuels generation	Ecosystems destruction	Yes	Local
Liquid					
Wastewater discharges	All the listed above	Coal and local fuels generation using a hydraulic ash sluicing	All the listed above	Partially	Local
Water consumption	–	Generation except DG units	Water supply deficit	No	Regional
Water heat pollution	Up to +30 °C	Heated water discharges	Local weather and ecosystems changes	No	Local

Table 2. Comparing old and modern conventional generation

Parameters	All kinds of old	PCB	HP GT and CCGT	GT	GP
SO_2, mg/m^3	500–800	100–200	~0	~0	~0
NO_x, mg/m^3	300–600	20–30	50–200	50–100	<500
CO, mg/m^3	~0	<20	50–200	60–200	<650
Particles, mg/m^3	<200	15–30	~0	~0	15–30
Efficiency, %	20–35	35–43	54–58	33–37	40–45
FUF, %	60–75	80–90	80–90	80–90	80–90

Given the absence of energy losses due to the small electrical distance to the customers' loads, distributed generation gives an advantage in efficiency of 5–15% and reduces the pollutant emission in 1.7–2.5 times.

Nevertheless, it should be taken into account that the hydrocarbon fuel distributed generation is inferior compared to the *new large-scale generation* by the environmental performance. This is concerned with the efficiency and the amount of the fuel burned (mostly for gas turbines), emissions of CO, NO_x and solid particles (mostly for gas piston units). It is associated with a scale effect as well as the constructive and technological features. Replacement of the existing large-scale generation by a great number of DG units instead of up-to-date large-scale generation leads to an increase in total pollutant emissions level. The same effect can be received if the load is implicitly covered by the DG instead the new large stations. DG allocation mainly depends on its stakeholders.

6. It should be noted that due to the properties of an area distribution and dispersion, the DG naturally reduces the concentration of the non-cumulative pollutants emitted. Due to economies of scale DG has a lower efficiency while CO_2 emissions and fuel consumption levels are higher, although DG development reduces a transmission and distribution losses due to its location at the point of energy consumption.

7. The additional DG advantage is, typically, the closed-loop cooling system without the necessity of water use and water heat pollution.

8. As for the reliability of DG compared to the large-scale generation, on the one hand, a multi-source power supply increases the structural reliability. On the other hand, the tendency of mostly gas fuel use for hydrocarbon DG decreases the reliability by economical factors, if the fuel is not diversified.

In general, mass development of hydrocabon DG can be effective only if replacing existing conventional generation, but not the new up-to-date large-scale one.

9. The direct effect on reducing emissions from reconstruction associated with replacing the outdated equipment by more efficient and environmentally friendly one. There is also an indirect effect that is estimated to be more than the direct effect. It is connected with the fact that the current economic and ecological benefits for the installed efficient equipment makes it competitive in the electricity and generation capacity markets compared to the less efficient generating equipment.

3 An example of the development planning

A real example of the successful development planning using the key points discussed above is presented. It is based on the industrial Sverdlovsk region of Russia where reducing power industry negative impact on the environment is urgent.

DOI: 10.1051/e3sconf/20160603011

The curves of pollutant emissions from power plants of the region are shown in Figure 1. The trend of wastewater discharges coupled with the water utilization by power plants is shown in Figure 2. The dynamics of waste disposal by power plants is shown in Figures 3.

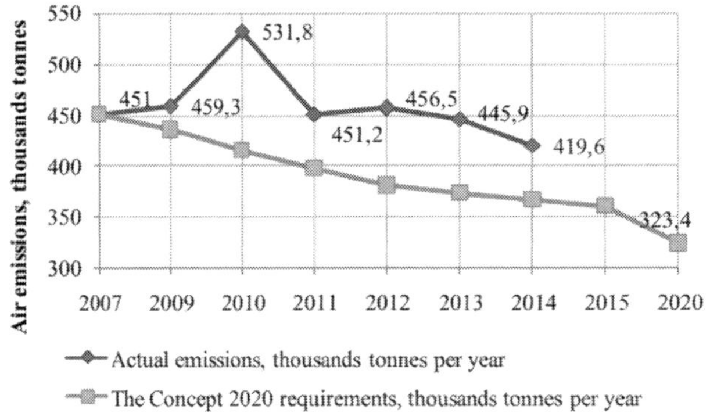

Fig. 1. Air emissions from power plants.

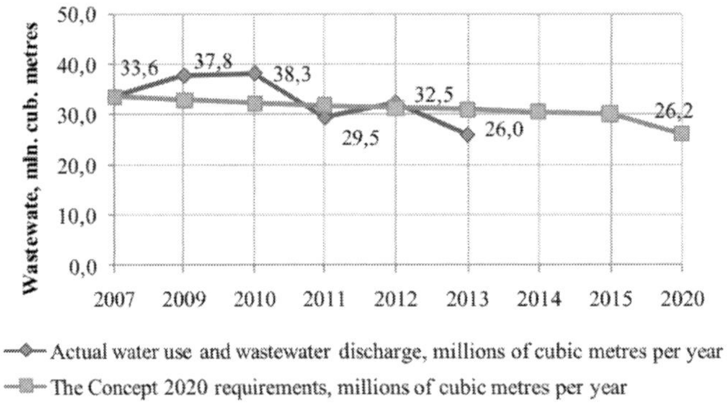

Fig. 2. Water utilization and pollution by power plants.

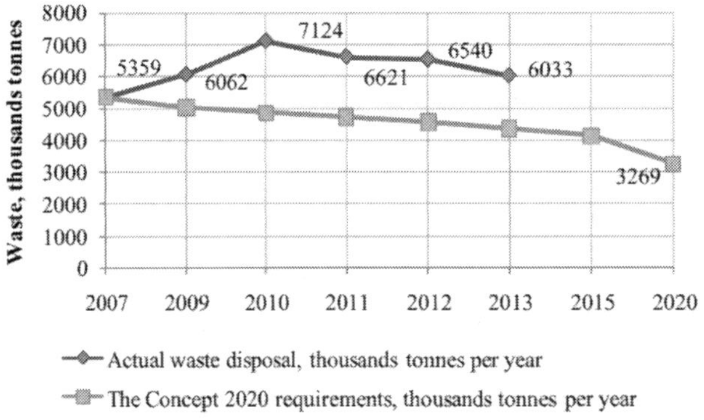

Fig. 3. Waste disposal by power plants.

The key points of sustainable development strategy, including energy efficiency and environmental impact aspects, are stated in the corresponding Conception document [5]. The Conception provides reducing the level of environmental load 2007–2020 to the values:

- pollutant emissions by 30%;

DOI: 10.1051/e3sconf/20160603011

- water utilization and wastewater discharge, by 30%;
- waste disposal by 40%.

As it can be seen from the figures 1–3, due to the industry development and energy consumption growth there was the corresponding growth of pollutant emissions, wastewater discharges and waste disposal from hydrocarbon power plants in the region. The actual curves sufficiently exceed the Concept ones; the peak was at the 2010. Gradual generation upgrade and reconstruction measures results in lowering the actual curves close to the planned values. Especially effective was waste utilization decrease due to the replacement of old steam-powered equipment by DG.

Also an example of the direct and indirect effects on emissions reduction is meaningful. Both effects are illustrated by Figure 4. The calculated reduction of emissions is made when replacing the existing equipment by the new one at the following power stations:
- Power Plant 1 – Verkhnetagilskaya Power Plant, 420 MW CCGT instead of 88 MW gas/carbon generation;
- Power Plant 2 – Serov, 420 MW CCGT unit instead of 388 MW carbon generation;
- Power Plant 3 – Nizhneturinsk, 420 MW CCGT unit instead of 176 MW gas/carbon generation.

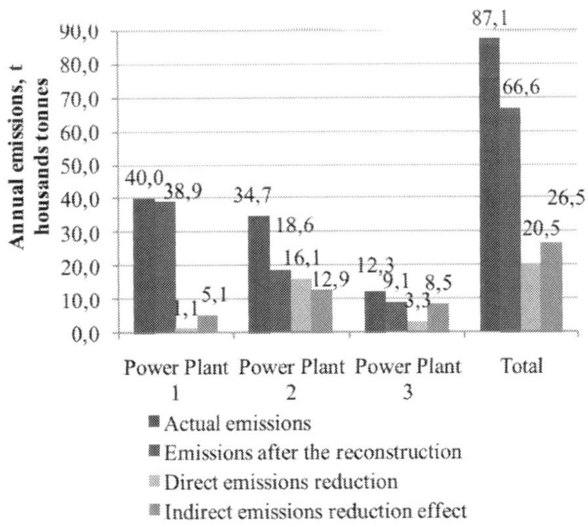

Fig. 4. Emissions reduction effects.

The total direct effect on the reconstruction of the station is estimated at 20 500 tons or 4.9% from the total emissions of power plants in the area (light green bars). The potential of the reduction by means of the market principles is 26 500 tons or 6.3% from the emissions of power plants in the area (dark green bars).

4 Conclusions

The following conclusion regarding the conventional sources upgrade can be made:

1. Conventional energy sources still play a predominant role in meeting power and energy balances, so, the sustainable development in a short-term and middle-term perspective to be implemented by the gradual conventional energy sources upgrade along with RES development.

2. In cities with high building density it is efficient to install gas generation based on CCGT and high-power gas turbines according to their level of emissions. Nevertheless,

the level and concentration of NOx to be accounted carefully when planning the development in terms of the strategy;

3. In small towns, suburban and industrial areas with a number of DG units based on gas turbines or gas piston engines can be installed. If a coal transport infrastructure and free land are available, it is possible to consider the modern modular carbon generating units equipped with dry ash removal and flue gas desulfurization as an alternative to gas generation. Similar properties have the power plants operating on local fuels.

4. In general, mass development of hydrocabon DG can be effective only if replacing existing conventional generation, but not the new up-to-date large-scale generation. DG allocation mainly depends on its stakeholders. This generation increases the total emissions in the area, but due to the properties of an area some distribution and dispersion, the DG naturally reduces the concentration of the non-cumulative pollutants emitted. Also DG may sufficiently reduce water use and water heat pollution.

5. The market-based indirect effect on emissions reduction from generation renovation can be more sufficient than the technical one.

Acknowledgement

The work was supported by Act 211 Government of the Russian Federation, contract № 02.A03.21.0006 and the Ministry of Education and Science of the Russian Federation (in the framework of state assignment, № 13.1928.2014/K (project №1928)).

References

1. Renewables 2015 Global Status Report. Renewable Energy Policy Network for the 21st century (2015)
2. Technology Roadmap: Nuclear Energy, OECD/IEA and OECD/NEA (2015)
3. The materials of public periodic seminar "The issues of distributed generation interconnection and operation". Russian National Committee of CIGRE. http://cigre.ru/activity/conference/seminar_c6
4. V.O. Samoylenko, and S.A. Eroshenko, The materials of public periodic seminar "The issues of distributed generation interconnection and operation", Russian National Committee of CIGRE, http://cigre.ru/activity/conference/seminar_c6/ materials/archive
5. The Concept of power industry development of the Sverdlovsk region 2016–2025. Ministry of power industry and housing and communal services (2015)

E3S Web of Conferences **6**, 03012 (2016) DOI: 10.1051/e3sconf/20160603012
ICSC 2016

Building an eco-effective district heating management system in a city

Leonid Gitelman[1], *Elena* Magaril[1], *Mikhail* Kozhevnikov[1a], *Giuseppe* Genon[2], *AnnaMaria* Senor[2], and *Debora* Panepinto[2]

[1] Research and Educational Centre ENGEC, Ural Federal University, 620002, Yekaterinburg, Russia
[2] Turin Polytechnic, 10129, Torino, Italy

Abstract. The goal of making the urban environment safer amid increasing human impact adds to the importance of district heating management. The article outlines the results of a study into the problem of improving the urban environment by implementing two innovative solutions. Technological innovations imply the introduction of modern sustainable tools of reducing emissions in district heating networks, one of them being the combination of district heating and combined heat and power plants (a case study of Turin). Organizational innovations are built upon the management of demand for thermal energy that makes it possible to reduce investment in new construction and to optimize the architecture of heat load schedules for the purpose of alleviating energy and environmental pressure on the city. The authors propose formats and areas of demand side management for thermal energy and methods of offering economic incentives to program participants.

1 Introduction

District heating networks in cities are characterized by inertia as regards their development, which is due to numerous factors affecting their relative competitive edge, such as the duration of the heating season, heat loads, growth in single-family construction; the structure of consumers. This is the reason why in different towns the share of consumers connected to district heating networks may vary greatly.

In developing countries, such as Russia, district heating networks are characterized by low economic efficiency and environmental effectiveness, primarily because of outdated and worn-out installations. Overhauling them would require huge investment. There is also a lack of interest from investors because of rigid tariff policies of governments and regulators.

In many countries, especially in big cities, heating is supplied by numerous organizations of various affiliations and ownership status (private, municipal, state-owned ones). At the same time, a unified integrated system for heat supply management in the city as a whole is practically non-existent. Such a system would coordinate the operation of these organizations and direct them towards meeting public interests, and create

[a] Corresponding author: np.fre@mail.ru

a competitive environment that would take into account the specific features of heating supply.

The context brings to the forefront the issue of municipal combined heat and power (CHP) plants. CHP plants have to simultaneously operate in electricity and heat markets that differ greatly in terms of their size, competition and demand. The efficiency of CHP plants during electricity generation is strongly dependent on heat loads, i.e., it is seasonal. Consequently, the ability of CHP plants to compete in the wholesale electricity market is objectively a variable. It is not ruled out though, that the generating capacity of some CHP plants will not be needed by the electricity market. The plants may turn into district boilers as a result, thus discrediting the idea of district heating in the long-term perspective.

The results of the study described by the authors focus on two elements of a heat supply management mechanism proposed by the authors. The mechanism includes economic, administrative and legal tools for serving the interest of major participants in the heat market, with the elements in focus being the improvement of environmental effectiveness and building a demand-side management system.

2 A study of environmental issues of heat supply in a city

An environmental analysis of the development of district heating (DH) was performed using the case of the DH system built in Turin (Piedmont, North Italy) and the forecast project of its expansion for the next eight years [1]. Technological solutions that are employed in the Turin DH system make it a highly interesting case to study in particular because they place the city at the core of the widest district heating network of Italy, which is also one of the most expansive DH systems in Europe. The system of combined heat and power production in Turin has proven its high efficiency in terms of energy savings, and makes it possible to minimize the environmental impact by integrating cogeneration into DH systems. The evaluation of the compatibility of the existing Turin plant configurations connected with DH may prove useful as a decision-making tool for policy-makers.

The research programme included the following steps.

1. Examination of social and economic factors that can potential become an obstacle to the development of district heating.
2. Building the energy and environmental balance of the city.
3. Evaluation of pollutant emissions (SO_2, NO_x, PM_{10}, CO_2) in cogeneration mode.
4. Building 3D models of pollutant dispersion.
5. Identifying the areas of the city with critically high concentrations of pollutants.
6. Optimization of the deployment of thermal power plants in the territory of study [2–5].

To define the dispersion and transport of pollutants, the AERMOD model was used, which is an evolution of the Gaussian model ISC3. The model was developed by the American Meteorological Society (AMS) for the Environmental Protection Agency (EPA). In order to establish ground level concentrations within an area of 40 x 40 km (Figure 1), the following methodology was used. First of all, a set of maps of ground-level concentrations following the activation of cogeneration units and their enlargement/reservation was formed. Then, for each zone of the district heating usage (domestic boilers) barycentric stacks were set. Finally, the result was obtained as the calculated difference between the added and the prevented (thanks to the replacement of thermal power plants) emissions.

The analysis leads to a number of conclusions:

- it was possible to establish the environmental compatibility of the CHP plants by assessing the possibility of ensuring significantly lower concentration values than allowed limits;
- information that is yielded by the emissive environmental balance is definitive only for CO_2;

• the ground-level concentration maps establish a fundamental baseline for the local analysis in the form of the mean annual values in the nearest municipalities.

In general, the obtained results highlight a sharp drop in the soil concentrations of nitrogen oxides and dust (PM10) that encourage the extension of the DH Scenario in place in Turin (Figure 2). On a global scale, net CO_2 reductions open up a path for the reduction of climate-altering emissions, as shown in Figure 3.

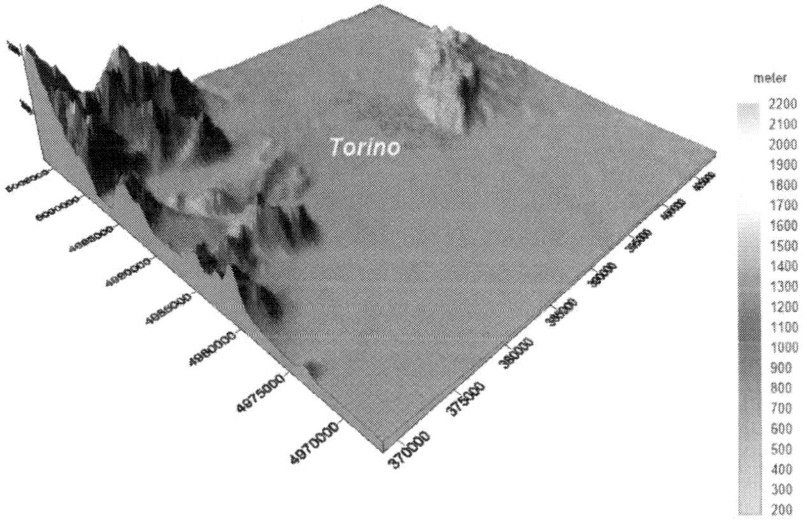

Fig. 1. Study Area.

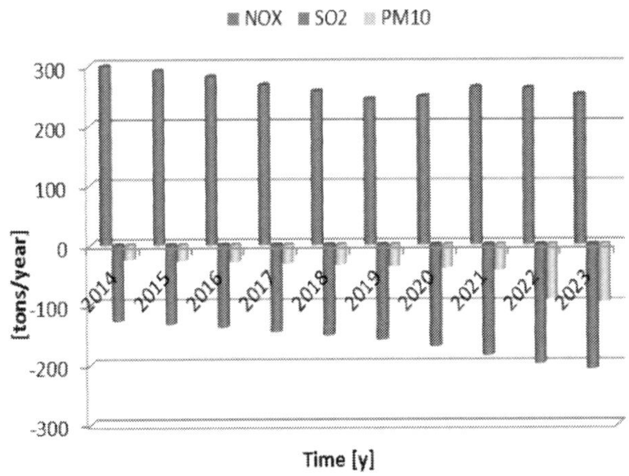

Fig. 2. Pollutant Local Balance.

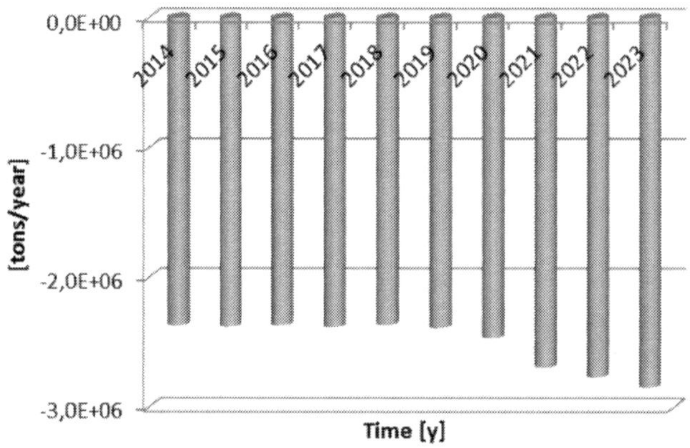

Fig. 3. CO_2 Global Balance.

3 Heat demand management

Thermal energy demand management targets end consumers (manufacturing companies, non-manufacturing organizations, population and others), specific energy consuming processes (technological, heating and cooling loads, hot water supply), various energy carriers (hot water and steam at temperatures and pressure). Within the framework of demand side management, the authors consider the following tasks:

- streamlining the operation schedule of CHP plants by combining an increase in power generation from waste heat and utilizing the flexibility potential of CHP plants during minimum daily and seasonal loads;
- reducing operating and capital costs thanks to the lower cost of recovered heat vs. the cost of heat generated by new installations, higher efficiency of CHP equipment and boilers and a reduction in energy losses in distribution networks and consumer units;
- improving the competitive edge of energy companies in heat markets through lower tariffs, a wider range of energy efficiency services and improved reliability of electricity supply (for example, by discouraging consumers from building their own heat generators);
- the elaboration of more substantiated plans for putting into service new generation capacity at CHP plants and boilers with lower investment risk;
- accelerated decommissioning of inefficient and eco-destructive energy installations;
- a radical improvement of business relations with customers, municipal and regional authorities by seeking a more effective alignment of commercial and public interests of all parties involved.

There is a number of formats and areas of heat demand management that could be proposed as a solution to the above tasks.

1. **Energy conservation.** It covers all categories of consumers and is implemented in an number of ways such as building, pipe and equipment insulation, installation of metering devices and heating controllers at homes, a wider use of renewable energy sources and installation of different energy conservation equipment at production companies, improved efficiency of heat consuming equipment and the use of advanced energy conservation technologies.

2. **Rational energy consumption scheduling.** It is aimed at smoothing out daily and seasonal load curves in industry (load management). It can take the following forms: peak reduction, valley filling, load shifting. It includes such organizational and technical

measures as shutting down equipment or load shedding during certain periods, a change in maintenance schedules, installation of standalone peak heat generators, steam accumulators etc.

3. **Substitution of energy carriers**. In principle, it is possible to substitute electric power for steam in some power-operated processes in production companies, and electric power for hot water in heating installations in the housing sector. It has to be emphasized that using CHP storage systems for heating purposes is not only a way to save thermal energy in the form of hot water, but also an effective way of filling the nighttime valley in the daily load curve. Heat demand management is, therefore, closely connected with electrification and electricity demand management [6]. There is an ensuing problem of striking a balance between the development of CHP units and separate heat and power production (steam-electric power stations and boilers). Obviously, as electricity consumption keeps growing, the share of CHP plants with low steam parameters and turbines with condensers will gradually go down because their operation in condensing mode is highly inefficient.

In low and medium temperature processes steam and hot water are also two competing energy carriers. Hot water can successfully replace steam not only for heating purposes, but also in some technological industrial process that use low-temperature heat..

4. **Reducing the parameters of energy carriers**. It acceptable to lower the pressure and temperature of process steam as well as of the temperature of hot water used for technological processes, heating and hot water supply in industrial facilities and community buildings, provided this does not disrupt production processes or affect the quality of products and minimum indoor thermal comfort. For example, water temperature in the heating systems of shopfloors could be lowered during working hours considering the heat output of equipment. For companies and organizations operating one and two shifts a day the air temperature could also be lowered significantly during night hours. Naturally, none of this is applicable to the housing sector.

Methods of providing economic incentives to DSM program participants are selected with regard to the specific features of the program, first and foremost the goals and subjects of demand side management. In the general case, the following system of methods and economic incentives could be considered.

1. Various ways of distributing cost savings in combined cycle generation (at CHP plants) between electric and thermal energy: those reducing the cost of electricity production; those reducing the cost of thermal energy, and those reducing the cost of thermal energy and electricity in comparison with separate heat and power production (a compromise solution).

The selection of a specific option is driven by the goals of incentives, such as substituting electric power for steam and hot water or increasing the competitive ability of an energy company in the heat market.

2. Tariffs for thermal energy that are differentiated by the energy carrier and its parameters (steam at various pressure levels, hot water), the type of the energy consuming process (technology, heating, hot water supply), time of consumption (time of the day, day of the week, season), and by the amount consumed.

To use tariffs as incentives, it is acceptable to allow them to differ from the actual cost of service (cost of energy supply). For example, in order to encourage energy conservation tariffs grow along with an increase in energy consumption (so-called progressive tariff). Special discounts and extra charges are applied to tariffs for competing energy carriers.

3. A one-time charge for connecting consumers to heating networks of the energy company, with its size being differentiated with regard to energy efficiency characteristics

of buildings to be connected. In this case additional revenue from the least efficient consumers will be used to fund discounts for the most efficient consumers.

4. Discounted tariffs for thermal energy for consumers who agree to intermittent restrictions on heat loads and the lowering of energy carrier parameters. A multi rate tariff with a separate charge for the connected heat capacity of the consumer seems to be the most appropriate option for this method. In this case, discounts are offered on the capacity fee (or the maximum declared load). Additionally, discounts can be offered on the price of energy efficient equipment, direct investment and targeted monetary loans for consumers.

4 Conclusion

When building a "green" district heating system in a city, a number of factors must be taken into account:

- technical characteristics of the existing systems (configuration, capacity, regulatory capabilities) and the structure of thermal energy generating capacity;
- the baseline level of reliability and quality of heat supply (deterioration, per unit cost of fuel, energy loss in mains and distribution networks, repair and maintenance costs);
- ground level pollutant concentrations in heating delivery areas;
- the structure of consumers that determines appropriate calculation methods and non-payment risk;
- the preparedness of municipalities to introduce the newest organization mechanisms of heat demand management and cooperation with energy companies and consumers.

It should be noted that it is increasingly important for energy companies to *become proactive* in creating demand rather than responding to it amid uncertainty about energy consumption, intensifying investment risk and stronger competition in energy markets and in heat markets in particular.

Acknowledgement

This research was supported by Act 211 Government of the Russian Federation, contract № 02.A03.21.0006.

References

1. A. Poggio, C. Maga, P. Benedetti, *Provincia di Torino. Piano di sviluppo del teleriscaldamento nell'area di Torino, Rapporto finale* (Tecnoapi, Torino, 2009)
2. A. Senor, D. Panepinto, G. Genon, WIT Trans. Built Env., **168**, 651–662 (2015)
3. M.F. Torchio, G. Genon, A. Poggio, M. Poggio, Energy, **34 (3)**, 220–227 (2009)
4. G. Genon, M.F. Torchio, A. Poggio, M. Poggio, Energ. Convers. Manage., **50 (3)**, 522–529 (2009)
5. D. Panepinto, A. Senor, G. Genon, WIT Trans. Ecol. Envir., **180**, 479–490 (2014)
6. L.D. Gitelman, B.E. Ratnikov, M.V. Kozhevnikov, Economy of Region, **2**, 78–84, (2013)

AUTHOR INDEX

EDP Sciences – Web of Conferences
17, Avenue du Hoggar
Parc d'Activité de Courtabouf
BP 112
F-91944 Les Ulis Cedex A
France

ISBN 978-1-5108-2606-9

9 781510 826069

CURRAN ASSOCIATES INC.
proceedings.com